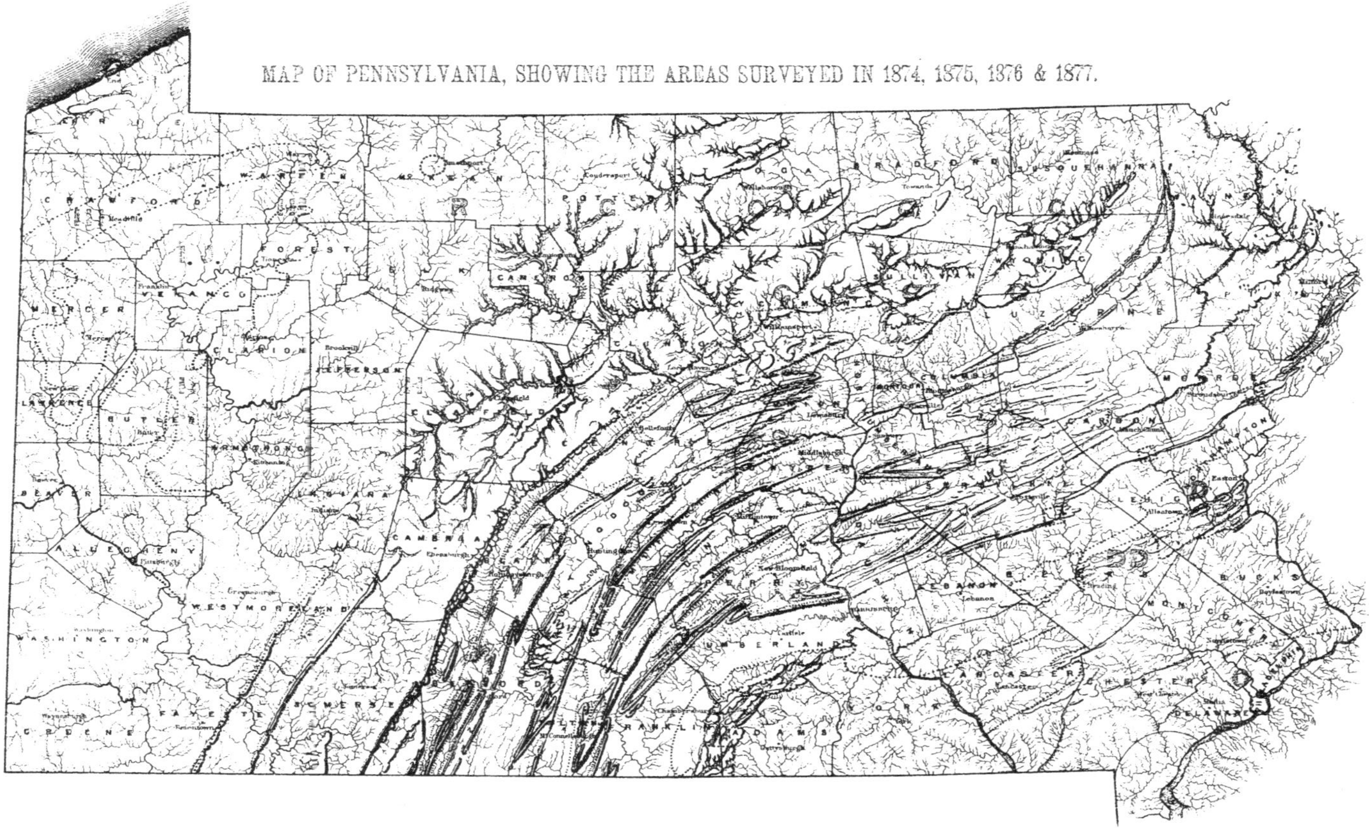

MAP OF PENNSYLVANIA, SHOWING THE AREAS SURVEYED IN 1874, 1875, 1876 & 1877.

SECOND GEOLOGICAL SURVEY OF PENNSYLVANIA:

1874.

PRELIMINARY REPORT

ON THE

MINERALOGY OF PENNSYLVANIA,

BY

F. A. GENTH.

WITH AN APPENDIX

ON THE

HYDROCARBON COMPOUNDS,

BY

SAMUEL P. SADTLER.

HARRISBURG:

PUBLISHED BY THE BOARD OF COMMISSIONERS

FOR THE SECOND GEOLOGICAL SURVEY.

1875.

LANE S. HART,
State Printer and Binder,

BOARD OF COMMISSIONERS.

His Excellency, JOHN F. HARTRANFT, *Governor*,
and *ex-officio* President of the Board. Harrisburg.

ARIO PARDEE, - - - - - - Hazleton.
WILLIAM A. INGHAM, - - - - Philadelphia.
HENRY S. ECKERT, - - - - - Reading.
HENRY M'CORMICK, - - - - - Harrisburg.
JAMES MACFARLANE,- - - - - Towanda.
JOHN B. PEARSE, - - - - Philadelphia.
ROBERT V. WILSON, M. D., - - Clearfield.
HON. DANIEL J. MORRELL, - - Johnstown.
HENRY W. OLIVER, - - - - - Pittsburg.
SAMUEL Q. BROWN, - - - - - Pleasantville.

SECRETARY OF THE BOARD.

JOHN B. PEARSE, - - - - - Philadelphia

STATE GEOLCGIST.

PETER LESLEY, - - - - - - Philadelphia.

LABORATORY OF THE UNIVERSITY OF PENNSYLVANIA,
Philadelphia, December 31, 1874.

PROF. J. P. LESLEY,

State Geologist:

SIR:—In obedience to instructions, I have prepared and have now the honor to submit the accompanying *Preliminary Report* on the Minerals found in the State of Pennsylvania.

The time which has elapsed since the organization of the Geological Survey of the State was so short that hardly anything more could be done than to give a list of localities where the different mineral species have been found. As the occurrence of minerals is often accidental, and comparatively very few only can be secured by a search at the locality, I had, besides visiting some of the most important mineral regions, to a great extent to depend upon the observations labors and the good will of others.

Pennsylvania being rich in private collections of great value, fortunately owned by friends and supporters of the geological survey, I have met with the most cordial reception everywhere, and was permitted to examine the specimens and take notes of their occurrences. Only by this generous assistance, which I have received from so many, I was able to accomplish my task.

To all who have aided me I wish to express my deep gratitude for their disinterested and very liberal support: to Mr. Wm. W. Jefferis, of West Chester, whose magnificent cabinet is very rich in many beautiful specimens from Lancaster, Chester and Delaware counties, the localities of many of which are now exhausted, and who has undoubtedly the best knowledge in the State of the mineral occurrences in this region, I am especially indebted for their revision.

I gratefully acknowledge the aid which I have received from Col. Joseph Willcox, of this city, who has paid special attention to the exploration of Delaware county, and who showed me the principal localities of this county, besides largely contributing from his fine cabinet, specimens for fuller investigation.

To Mr. Theodore D. Rand, who for many years has made the mineralogy of the neighborhood of Philadelphia a careful study, and who is better informed on this subject than anybody else I am under great obligations not only for a complete list of the localities of this vicinity; but also for specimens for examination, which he has presented me from his cabinet, so rich in local occurrences, and for visiting with me some of the principal localities near this city.

Prof. W. Theodore Roepper of Bethlehem, the accomplished mineralogist and analytical chemist, not only showed me the occurrences in the neighborhood of Bethlehem, especially in Lehigh and Northampton counties and gave me the localities of many others, but he liberally contributed very valuable analytical results, not yet published.

Mr. Samuel Tyson, of King of Prussia, furnished me with a list of Pennsylvania minerals in his fine cabinet, and was the first to call my attention to the interesting locality of zeolites at Fritz's Island, of which he presented me with good specimens; a still larger number and the best which have been found were very liberally supplied by Mr. H. W. Hollenbush, of Reading, who also sent me a list of Berks county localities.

Many thanks are due also Mr. Joseph Wharton, of Philadelphia, to Dr. John M. Cardesa, of Claymont, Delaware, to Dr. George Smith, of Upper Darby, and Mr. Lewis Palmer, of Media, Delaware county, to Mr. C. M. Wheatley, of Phœnixville, Mr. J. Taylor Boyd, of Cornwall, Lebanon county, and to many others.

To Mr. E. Mortimer Bye, of Wilmington, Delaware, I am especially indebted for very valuable services which he has rendered in accompanying me to the principal chrome ore localities of Lancaster and Chester counties, and for the statistics of chrome ore operations.

The main object of this preliminary report being to prepare as complete a list of mineral occurrences in Pennsylvania as the information which we possess at present, permits, I have endeavored to make this as full as possible, omitting only such as are on account of their wide distribution found almost everywhere, and such as form constituents of rock-masses, and which therefore occur wherever the latter are found. The number of localities, if not species, will undoubtedly be much increased when the results of the exploration of the geologists in the field become known to me, and when districts are explored which as yet have not received the deserved attention.

When a final mineralogy of the State is prepared, which should contain not only all the species and varieties occurring in Pennsylvania, but also a careful study of their crystalline forms, illustrated by drawings, &c., and an investigation of their optical properties and chemical analyses of all that is new and important, a more complete and systematic list of localities should be attached. To make this what it should be requires the assistance of all friends of mineralogy, whom I hereby request to favor me not only with any suggestions or corrections which they may have to make in this report, but also with any new observations and occurrences which they may have found, together with specimens from the same.—As a great favor I would consider the communication of everything that is new or not fully investigated.

In the arrangement of the species I have followed that of Dana's System of Mineralogy, fifth edition, 1868, on account of the excellence of this work. It is the most universally adopted and therefore in the hands of all who take an interest in mineralogy, and the arrangement of the species is the most natural one, grouping those of the same chemical constitution together, and sub-dividing them according to their forms.

Before giving the localities of every species which occurs in the State, I have given for the more important ones the distinguishing characters in a few words, in most cases with very little alteration, often verbatim, taken from this work; as the time was too short to pay any attention to the crystallographic and optical properties of the minerals, a discussion of these has to a great extent been omitted and reserved for future reports. More attention has been paid to their chemical composition. All the analyses, which I have been able to find of Pennsylvania minerals have been given, excepting a few which are obviously erroneous; some of the old and less perfect analyses were also introduced when they were interesting in a historic point of view, or of importance in showing the correctness of the views advocated in this report.

In the Report on the Geology of Pennsylvania, by Prof. H. D. Rogers (1858), numerous analyses of iron ores, limestones, coals, &c., are published, the results of very valuable labors of Prof. R. E. Rogers. To insert these would have very much augmented my report; many of the analyses are very interesting, especially those of limestones, and prove the character and position of the true and dolomitic varieties. I give only a few new analyses of these, which have been made this year in the Laboratory of the University of Pennsylvania.

The older analyses of the iron ores to be of value for practical purposes are wanting in a very important element, viz.: a *correct determination of phosphoric acid.*

Still a matter of great difficulty and requiring much experience, there was at the time when these analyses were made hardly a method known which gave reliable results; and the necessity of an accurate determination of phosphoric acid was not then appreciated.

Of several doubtful species or interesting varieties *new* analyses were made by myself or under my control by students in the University of Pennsylvania.

An investigation of the chemical and mineralogical constitution of rock-masses has been commenced, but very little progress has been made in it yet; a few of the results obtained I give under the head of the principal mineral constituent. Such an investigation, which would embrace the optical examination of thinly ground sections of the rocks, and careful analyses of their soluble and insoluble constituents, will be the only means to ascertain the nature of our numerous trap dykes and other crypto-crystalline rocks.

The few reliable analyses of Mineral waters from Pennsylvania which I have been able to find, I give under the heading of Halite or Common Salt, predominantly contained in all. With regard to minerals, resulting from decomposition or decay of organic substances—anthracite, semi-bituminous coal, bituminous coal, lignite and petroleum—a short review will be given by Prof. Samuel P. Sadtler, of the University of Pennsylvania, and will be appended to this report. For analyses of the coals, &c., readers will be referred to H. D. Rogers' Geology of Pennsylvania. Of much interest and importance will be the article on Petroleum, showing all that has been done in the investigation of this important substance and especially how little we know of this subject and how much is yet to be learned by fuller investigation.

Such an investigation could not be placed in better hands than Dr. Sadtler's.

I close with the hope that this "Preliminary Report on the Minerals of Pennsylvania" may receive your approbation. No one can be more fully aware of the many imperfections and short-comings in the same than I am, but I trust that the new facts and some of the suggestions therein contained may outweigh its defects.

Very respectfully,

Your obedient servant,

F. A. GENTH.

CLASSIFICATION.

The general subdivisions in the classification of minerals adopted in Dana's System of Mineralogy, has been followed in this Report.

It is as follows:

I. Native Elements.

II. Compounds.

A. *Binary:* Sulphides, Tellurides of metals of the Sulphur and Arsenic groups;

B. *Binary:* Sulphides, Tellurides, Selenides, Arsenides, Antimonides, Bismuthides of metals of the gold, iron and tin groups;

C. *Ternary:* Sulpharsenides, Sulphantimonides, Sulphbismuthides.

III. Compounds.

Chlorides, Bromides, Iodides.

IV. Compounds.

Fluorides.

V. Compounds.

Oxygen compounds.

A. *Binary: Oxides.*

B. *Ternary:* 1. Silicates; 2. Columbates and Tantalates; 3. Phosphates, Arsenates, Antimonates, Vanadates,* Nitrates; 4. Borates; 5. Tungstates and Molybdates; 6. Sulphates, Chromates, Tellurates; 7. Carbonates; 8. Oxalates.

VI. Hydrocarbon Compounds. *Minerals of organic origin.*

* I have placed the Vanadates to the Phosphates, &c., where they properly belong.

CHAPTER I.

NATIVE ELEMENTS.

1. *Gold.*

Notwithstanding the numerous reports, which from time to time have appeared, of the discovery of rich deposits of precious metals in Pennsylvania, there are only a few authentic localities at which small quantities have been found.

The physical properties of gold are too well known to require enumeration.

Dr. Ch. M. Wetherill (Transactions Am. Phil. Soc., X, 350), not only observed traces of gold in ferruginous quartz and pyrite on the land of Mr. Yoder, in Franconia township, Montgomery county, but also on the same property, fine *gold* spangles in the gravel and sand thrown out while digging a well. He found the quantity of gold in this gravel equal to 26½ cents in 100 lbs.

A ferruginous quartz and pyrite from Penn's Mount, behind Reading, also contained, according to his authority, traces of gold.

Traces of gold I have found in the chalcopyrite or copper pyrites from the Chester County Mines, near Phœnixville, and the ores from the Gap Mine.

One of the workmen in Mr. Charles Lennig's Chemical Works, by the name of Kuhbach (Keller-Tiedemann's Nordamerikanischer Monatsbericht I, 231; also, F. Fraley, in Proc. Am. Phil. Soc., V, 313), who was familiar with the operations of washing gold on the Rhine, found in the gravel of the Delaware River, at Bridesburg, *native gold* in scales, accompanied by menaccanite, magnetite, etc. This discovery was verified by Mr. Wm. Schrader, then chemist at the works, who extracted the gold from the sands by mercury, and reduced it to a pure state. The quantity of gold which a hand could wash from the Delaware River sands, was variously estimated at from 25 to 60 cents per day.

An exceedingly interesting account of the occurrence of gold in Pennsylvania, is given in a paper by Messrs. Dubois & Eckfeldt (Proc. Am. Phil. Soc., VIII, 273), " on the Natural Dissemination of Gold."

An assay of the galenite of New Britain, in Bucks county, showed the presence of 2¼ grains, or not quite 10 cents worth of gold per ton. The speck of gold, which they have extracted from five ounces of galenite from this locality, is exhibited in the cabinet of the U. S. Mint in Philadelphia.

In the description of the occurrence of gold in the clays underlying the city of Philadelphia, I use their own language.

"Perhaps the most curious result of all is, that which remains to be stated.

"Underneath the paved city of Philadelphia, there lies a deposit of clay, whose area, by a probable estimate, would measure over three miles square, enabling us to figure out the convenient sum of ten square miles. The average depth is believed to be not less than fifteen feet. The inquiry was started whether gold was diffused in this earthy bed. From a central locality which might afford a fair assay for the whole, the cellar of the new market house in Market street near Eleventh street, we dug out some of the clay at a depth of fourteen feet, where it could not have been an artificial deposit. The weight of 130 grammes was dried and duly treated, and yielded one-eighth of a milligramme of gold, a very decided quantity on a fine assay balance. It was afterwards ascertained that the clay in its natural moisture loses about fifteen per cent by drying. So that as it lies in the ground the clay contains one part in 1,224,000.

"This experiment was repeated upon clay taken from a brick yard in the suburbs of the city, with nearly the same result.

"In order to calculate with some accuracy the value of this body of wealth, we cut out blocks of the clay, and found that on an average, a cubic foot, as it lies in the ground, weighs 120 pounds, as near as may be; making the specific gravity 1.92. The assay gives seven-tenths of a grain, say three cents worth of gold to the cubic foot. Assuming the data already given, we get 4180 millions of cubic feet of clay under our streets and houses, in which securely lies 126 millions of dollars. And if, as is pretty certain, the corporate limits of the city would afford eight times this bulk of clay, we have more gold than has yet (1861) been brought, according to the statistics, from California and Australia."

Still more astonishing—but, unfortunately, equally devoid of practical value—would have been the results, if the gravel which *underlies* this auriferous clay, and which is always richer than the clay above it, had been examined for gold.

2. *Silver.*

Silver is mentioned as having been observed in its *native* state at the Wheatley Mines, near Phœnixville, Chester county (Report on the Wheatley, Brookdale and Charlestown Mines by Prof. Henry D. Rogers, May 1, 1853, and his Geology of Penna., II, 701).

The so-called "gossans" from the same mines also contain silver, probably in the native state, in quantities varying from ¼ to 10 ounces per ton of ore.

The greater part of the silver is found in the form of argentiferous galenite, smaller quantities occur in the carbonates and phosphates of lead.

From numerous assays of argentiferous galenites from Pennsylvania which I have made, it appears that those from the Pequea Mines in Lancaster county, will yield on an average 250 to 300 ounces per ton.

Chester county veins near Phœnixville, viz., Chester county, Wheatley, Brookdale mines, etc., 10 to 40 ounces per ton.
Ecton Mine, Montgomery county, 5 to 10 ounces per ton.
New Britain Mine, Bucks county, 10 to 15 ounces per ton

3. *Copper.*

Beautiful specimens of native copper have been observed in Pennsylvania.

The finest come from the great magnetic iron ore mines at Cornwall, Lebanon county.

The crystals are mostly cubes with octahedral and dodecahedral planes, but usually so much distorted that it is almost impossible to make out their form. They are sometimes arranged in arborescent groups or in reticulated sheets covered with brilliant crystals, sometimes in crystalline plates. The crystals are bright copper red, but occasionally coated with a film of cuprite which gives them a dull red appearance. They are found between the magnetite or on quartz.

Native copper has also been found under similar circumstances at the Jones' Iron Mines, Berks county, and the French Creek Mines, near Knauertown, Chester county.

It occurs sparingly in delicate films on limonite, or quartz crystals, or as an interposing layer between limonite and chalcopyrite at the mines near Phœnixville and at Perkiomen, also in sheets of arborescent crystals, at the Gap Mine, Lancaster county, and in the epidote rock at several localities in Adams county, for instance at Fairfield, at Maria Furnace, and just south of Gettysburg.

An interesting occurrence has been noticed by Prof. W. Th. Roepper in the limonite deposits of Ironton, Lehigh county, where it is occasionally met with in minute distorted, probably cubical crystals disseminated through quartz.

In no locality in Pennsylvania, has native copper been found in sufficient quantity to be valuable as an ore.

4. *Iron.*

Iron in its *native* state has not yet been observed in Pennsylvania; but as the trap rocks of New Jersey contain it, according to Prof. G. H. Cook's observation, which he communicated to me, it is more than probable that it will be found in this State.

I have to mention here two masses of iron, which are supposed to be of meteoric origin.

a. Bedford County Iron, 1828.

Prof. C. U. Shepard describes it (Am. Jour. of Sc., XIV, 183), as "*Native Iron? slightly arseniuretted*,"—a fragment of two to three ounces in weight. It had evidently formed distinct crystals—was highly crystalline and apparently rhombic; it shows perfect cleavages, giving angles approximately of 121° and 59° for the primary planes and 146° for the secondary planes, intersecting the base parallel to its greater diagonal. The cleavage parallel to the lateral planes is easily effected, whilst no terminal one is visible, it breaking in that direction with great difficulty and presenting a sub-hackly fracture. It has a fine metallic lustre, and a color between silver-white and steel-gray. It breaks with much difficulty and small masses flatten out like pure iron. Its hardness is almost equal to that of ordinary steel. Highly magnetic with polarity, sp. gr.=7.337. Before the compound blowpipe, emits the odor of arsenic. Dissolved in nitric acid, it yields little black flakes which were found to be graphitic carbon.

The analysis gave:

Iron	=	97.44
Arsenic	=	1.56
Graphite	=	0.40
Loss	=	0.60
		100.00

In a subsequent "Report on Meteorites," by C. U. Shepard (Am. Jour. of Sc., [2] II, 391 and IV, 85), he places it together with the Randolph county, (N. C.) Iron, which it resembles in structure, color, hardness and lustre, its sp. gr.=6.915, and says that he had been unsuccessful in verifying the existence of arsenic or of detecting any other metal, but that "its greater analogy to the Randolph Iron than to any other terrestrial production, either natural or artificial, induces me to retain it in the category of meteorites."

In his "Catalogue of Meteorites" (Am. Jour. of Sc., [2] XXXI, 459), he places it amongst the "numerous native iron and steel-like masses, whose origin is not understood, but which are suspected to be meteoric."*

It appears to be only a piece of "spiegeleisen." The analysis given is evidently erroneous.

b. Pittsburg Iron, 1850.

It was described by Prof. Benjamin Silliman, Jr., at the meeting of the American Association for the Advancement of Science, held at New Haven, in August, 1850, and a short description of it published in their Proceedings, page 37. Mr. John H. Bailey, of Pittsburg, gives an account of it in a letter to Prof. Silliman, dated June, 1850, from which it appears "that

* See No. 23, Leucopyrite.

this mass of meteoric iron was found in a field upon Miller's Run, in Alleghany county, near Pittsburg. A farmer was ploughing in the field, where, seeing a snake, he seized a stone, as he supposed, to destroy the animal, but finding it remarkably heavy, he was attracted, after completing his purpose, to examine the body which possessed such a remarkable weight. It was carried to Pittsburg, where it was found to be very malleable, but unfortunately wrought into a bar, which has since been lost sight of. The mass was of an ovoidal figure, almost six or seven inches in diameter and weighed nearly 292 (?) pounds. It is greatly to be regretted that only a very small portion of this large mass has been preserved. A qualitative examination has shown that it is rich in nickel and possesses only a very inconsiderable portion insoluble in acids."

A fuller analysis has, as far as I can learn, never been made. Its sp. gr. = 7.380 (C. U. Shepard, Am. Jour. Sc., [2] XI, 40).

5. *Tin.*

This very rare mineral has been observed by Dr. Charles M. Wetherill (Trans. Am. Phil. Soc., X, 350), who states that it occurs with gold in gravel of Franconia township, Montgomery county, the largest pieces adhering to the gravel and forming a rounded mass of a *white malleable metal,* which by analysis was proved to be *tin.* By panning, an additional quantity of spangles of native tin were obtained.

6. *Sulphur.*

It is found in cavities of galenite, in minute rhombic, highly modified octahedra of brilliant pale greenish yellow color, and resulting from its alteration, at the Wheatley Mine, near Phœnixville.

Also in minute crystals at the Burning Mine at Summit Hill, Carbon county; it has been occasionally met with in several of the coal mines as a pulverulent coating.

At Barren Hill, Montgomery county, it has been found as a granular powder in pyritous quartz.

7. *Graphite* (*Werner*).

Graphite or plumbago crystallizes in the hexagonal system, mostly in flat, six-sided tables, with eminent basal cleavage. Usually in foliated or granular masses; also impalpable. H = 1—2; sp. gr. = 2, 1—2.23; color iron-black to dark steel-gray. Lustre, metallic. Rarely found pure, and then consisting of pure carbon; often mixed with oxide of iron, quartz, mica, clay, &c.

Occurs in delicate hexagonal tables of an iron-black color and metallic lustre, disseminated through other minerals, also massive, at George Van Arsdale's quarry, near Feisterville, Bucks county.

A mine of very pure plumbago has been worked, over one hundred years ago, near Bustleton, Bucks county.

An analysis of the latter by Lardner Vanuxem (Jour. Acad. Nat. Sciences, Phil., V, 17), gave the following results:

Carbon	=	94.40
Silex	=	2 60
Oxide of iron and manganese	=	1 40
Water	=	0.60
Loss	=	1.00
		100.00

Granular and foliated masses have been found imbedded in talc and tremolite at Chestnut Hill, near Easton; small scales in the crystalline limestone, three miles north of Bethlehem, on the Monocacy creek; in magnetite at Siessholtzville, Lehigh county; in limonite at Yellow Springs, Chester county; in blueish quartz in Pikeland, Chester county; in quartz in East Bradford, Charleston and Uwchlan townships, Chester county.

A massive variety has been observed at Robinson's Hill, on the Schuylkill, five miles from Philadelphia, and an impure pulverulent variety, much mixed with foreign matter, at Pughtown, S. Coventry, Chester county. The latter has been used as a mineral paint; it is also found two miles northeast of Jones Mine, in Berks county, and several other localities.

A partial analysis of samples from S. Coventry gave me 7.20 per cent., and one from Berks county 10.85 per cent. of carbon.

I believe that it will be interesting to mention that the gray color of the roofing slates at Peachbottom, in York county, also ot those in Lancaster county, is due to about 0.5 per cent. of carbon in its graphitic form.

CHAPTER II.

COMPOUNDS.

A. BINARY.

Sulphides, &c., of metals of the Sulphur and Arsenic Groups.

8. *Stibnite* (*Dana*).

Mr. Samuel Tyson, of King of Prussia, made the interesting observation that this mineral occurs with the zeolites of Fritz's Island, near Reading. He found only a few very minute crystals which were imperfect rhombic prisms longitudinally striated and with eminent brachydiagonal cleavage, and of a dark lead gray color. A blowpipe examination which I made proved them to be tersulphide of antimony or stibnite.

9. *Bismuthinite* (*Dana*).

Very small quantities of the tersulphide of bismuth have been found by Mr. Theodore D. Rand, in tourmaline, in a granite vein in gneiss in West Philadelphia, opposite the Fairmount Water-works.

10. *Molybdenite* (*Brongniart*).

It occurs in tabular hexagonal plates, usually foliated, massive, or in fine scales, with an eminent basal cleavage. H. = 1—1.2; sp. gr. = 4,4—4, 8. Pure lead gray color and metallic lustre. Its composition is $Mo\ S_2$ = Molybdenum 59.0, Sulphur 41.0.

It occurs in beautiful hexagonal tabular crystals and small foliated masses of a lead gray color and brilliant metallic lustre in gneiss at Frankford; also in crystals and foliated masses in quartz, near Chester, Delaware county; at Zion's Church, and on the farm of V. Hartman, in Alsace township, near Reading, Berks county. In small laminated patches in gneiss, it is said to have been found at Fairmount, Philadelphia.

The molybdenite from Chester has been analysed by Henry Seybert (Am. Jour. Sc., IV, 320), and that from Zion's Church, near Reading, by Dr. Charles M. Wetherill (Trans. Phil. Soc. X, 345).

		Chester, Del. Co.	Zion's Church, near Reading.		
Sp. gr.	=	4.444	——		deducting impurities.
Molybdenum	=	59.42	55.727	=	59.332
Sulphur	=	39.68	38.198	=	40.668
		99.10	Silicic acid =2.283		100.00
			Ferric oxide=3.495		
			Water =0.297		
			100.000		

Dana's Mineralogy gives George Van Arsdale's quarry in Bucks county, as a locality. This is doubtful; all that I have seen from there, resembling molybdenite, was graphite.

B. Binary.

Sulphides, &c., of metals of the gold, iron and tin groups.

11. *Galenite (von Kobell).*

It crystallizes in the isometric system; the most common forms are cubes in combination with octahedral and dodecahedral planes. Cleavage eminently cubical, rarely octahedral. Also coarse and fine granular, fibrous and sometimes impalpable. H = 2.5—2.75; sp. gr. = 7.25—7.7. Color pure lead gray. The composition is sulphide of lead, PbS = lead 86.6 Sulphur 13.4. Frequently contains minute quantities of silver, copper, antimony and other metals.

This mineral forms veins of considerable size and richness in the gneiss in the neighborhood of Phœnixville, Chester county, which have been worked to some extent and with encouraging results at the Wheatley, Brookdale, Chester county and other mines. The veins sometimes penetrate into the Triassic sandstones, or are entirely contained in these, as at the Ecton Mine, Montgomery county. Near the surface the galenite is to a considerable extent converted into a beautiful variety of lead salts, of which I shall speak in the sequel.

At these mines near Phœnixville, galenite is rarely met with in crystals, however some very fine cubes have been observed, and cubes with octahedral and dodecahedral planes. The crystals are often rounded, almost globular. The greater portion of it is coarsely granular, sometimes also fine-grained and fibrous. As already stated, it contains from 10 to 40 ounces of silver per ton of ore.

According to Prof. H. D. Roger's Report (*l. c.*) an antimonial sulphide of lead and silver (probably only a variety of galenite) occurs at the Wheatley Mine.

The galenite from the Ecton Mine, Montgomery county, is rarely found in cubooctahedra, mostly the coarse granular but sometimes also in the fine-grained variety.

A similar variety has been discovered and somewhat explored at New Britain, in Bucks county.

Small quantities of galenite accompany the zinc ores of Sinking Valley, in Blair county; those in the neighborhood of and five miles south of Lancaster and four miles northwest of Lancaster.

Minute quantities associated with pyrite occur occasionally in the carboniferous sandstone of Bradford county, and near Pottsville, in the shaft of the Philadelphia and Reading Railroad Company.

It has been observed associated with crystallized blende in limestone at Espy, near Bloomsburg and four miles south of Bainbridge on the Susquehanna; and Prof. Lesley has found a cube of galenite surrounded by clay in argillaceous siderite near Pittsburg.

The galenite of the Pequea Mine in Lancaster county is very interesting.

The coarse granular variety containing only a small quantity of silver occurs in a small irregular vein in the mica slate; there is another variety, however, which, either alone or associated with quartz, feldspar and mica—a regular granite—forms small lenticular, vein-like segregations in the lower Silurian limestone. It generally shows a very perfect cubical cleavage and is highly argentiferous, yielding as already stated from 250 to 300 ounces of silver per ton.

Besides this variety with cubical cleavage there is another having a distinct octahedral cleavage into which the first sometimes graduates. It was first noticed by the late Dr. John Torrey, who described it in a letter to Prof. G. J. Brush (Am. Journ. of Sc. [2] XXXV, 126), however erroneously stating the locality as being in Lebanon county. The cleavage is, in many of the pieces eminently octahedral, the cubical being less distinct; on heating, however, the latter becomes more predominant.

The spec. gr. was found to be 7.63. The amount of silver found by Dr. Torrey was 179⅓ ounces per ton.

The fact of presenting an eminent cleavage parallel to the octahedral planes has given rise to several suggestions and interesting experiments by Prof. Josiah P. Cooke, Jr. Dr. Torrey was of opinion that this variety may be a pseudomorph after fluorite, or that it may prove the galenite to be dimorphous or crystallizing in two distinct forms. Professor Cooke showed that although the cubical cleavage of galenite is the usual, that which is parallel to the octahedron can be obtained *by pressure* with many galenites from other localities, and suggests that the Pequea galenite may be the result of the latter.

Interesting as the observation is, I do not believe that it requires any other but the most simple explanation, viz.: that the individual particles, which constitute the mass of galenite, have crystallized octahedrally (a very frequent form), which in their aggregate show a predominance of the original form over the cleavage, and therefore a more ready separation in this direction. A peculiar film and roughness of the octahedral planes speak in favor of this view.

These octahedral galenites are certainly neither pseudomorphs nor a dimorphous form of sulphide of lead.

12. *Bornite* (*Haidinger*).

Bornite also called erubescite or variegated copper ore, crystallizes in the isometric system; good crystals are rare, and none have been observed in this State. Usually in compact or granular masses. H = 3; sp. gr. = 4.4—5.5. Color pinchbeck brown inclining to copper-red, readily tarnishing with purplish colors. Variable in its chemical composition, probably on account of admixtures of chalcocite or chalcopyrite. The pure bornite is: $3\ Cu_2\ S + Fe_2\ S_3$, containing copper 55.58, iron 16.37, sulphur 28.05.

This very valuable copper ore has occasionally been met with at different localities in the State, but nowhere as yet in sufficient quantities for practical purposes.

The largest deposit is probably found in the Triassic rocks of York county; it has also occurred near Gettysburg in Adams county; at St. Mary's in Warwick township, Chester county; in the gneiss rocks at Frankford, and at McKinney's quarry, near Germantown, and with chlorite in the soapstone quarries near Lafayette, Montgomery county.

It is also found in the copper horizon of the upper Devonian (Catskill Group) in Lycoming, Sullivan and other counties.

13. *Pentlandite* (?) (*Dufrenoy*).

This is an isometric non-magnetic mineral with octahedral cleavage, a light bronze color and the composition R S (R being Ni and Fe). The analysis by Scheerer (Pogg. Ann. LVIII, 315) agreeing closely with NiS + 2FeS = Nickel 22.1, iron 41.9, sulphur 36.00.

I have observed that a small portion, perhaps not over 5 or 6 per cent. of the niccoliferous pyrrhotite from the Gap Mine, Lancaster county (see below) is not attracted by the magnet. I therefore have requested Mr. Henry Pemberton, Jr., of the Laboratory of the University of Pennsylvania to make an analysis of the *non-magnetic* portion, in which he found as follows:

Insoluble hornblendic mineral	=	24.81
Magnesia	=	2.73
Sulphur	=	23.56
Iron	=	22.22
Nickel	=	25.73
		99.05

Calculating from these results the composition of the mixture, I find it to be:

Sulphide of nickel	=	39.96
Sulphide of iron	=	26.40
Ferrous oxide	=	6.00
Magnesia	=	2.73
Hornblende, silica, &c.	=	24.81
		99.90

The ratio between the sulphides of nickel and iron is 3 NiS + 2FeS, giving the following composition of the pure mineral.

Sulphur	=	35.64
Nickel	=	39.42
Iron	=	24.94
		100.00

Although this composition is, in so far, different from pentlandite—that the Gap Mine mineral contains a far larger percentage of nickel and much less iron—the atomic ratio of the combined metals remains the same, and for this reason I place it under pentlandite until it can be obtained in a *pure* state for a fuller investigation.

14. *Sphalerite* (*Glocker*).

Sphalerite, or as it is usually called *zincblende* or *blende*, crystallizes in isometric tetrahedral crystals often highly modified and complicated, with perfect dodecahedral cleavage; frequently in twins; also massive, granular, compact. H = 3.5 — 4; sp. gr = 3.9—4.2; colors white, yellow, yellowish and grayish white, brown, black. Lustre resinous to adamantine. In its pure state, sulphide of zinc, ZnS = Zinc 67. Sulphur 33. Frequently part of the zinc replaced by iron or cadmium

This mineral is found in the State not only in beautiful specimens but also in large masses, forming a very valuable ore.

The most beautiful crystals of a pale wine yellow and greenish yellow color, some in almost perfect dodecahedra from ¼ to 1 inch in diameter and smaller crystals in more complicated forms, have been found at the so-called Napoleon Mine, Valley Forge, Montgomery county. Yellow blende has also been found at the Phœnixville Tunnel, and in calcite, at Webb's Mine, Columbia county.

Yellow, brown and black varieties—sometimes crystallized but generally in masses showing a highly perfect dodecahedral cleavage, occur associated with galenite at the Ecton Mine, near Shannonville, Montgomery county.

A considerable quantity of this blende has been mined and sent to market.

A very similar zincblende of a pale brown color, often in beautiful and large crystals, mostly modified dodecahedra, and twins has been found at the Wheatley, Brookdale, Chester county, &c , Mines, near Phœnixville.

The large quantity of this mineral at these mines, was at the time these were in operation a very serious detriment, because no market could be found for the same in this country. Small quantities of it from the Wheatley Mines were exported to England.

The largest deposits of zincblende occur at the great zinc mines near Friedensville, in the Saucon Valley, Lehigh county. It is rarely met with in minute yellow crystals, more frequently fine-grained or crypto-crystalline, massive and compact, and of yellowish and grayish white colors, and a waxy lustre. It is a very important ore of zinc, and now very largely reduced both into the white zincoxide and into metal.

Smaller quantities of zincblende occur with the galenite at New Britain, in Bucks county, and with the zinc ores of the Sinking Valley Mines, in Blair county, and, frequently mixed with dolomites, at the Lancaster Zinc Mines, 5 miles south of Lancaster, and at the Pequea Mine, Lancaster county.

Also in the limestone at Espy, near Bloomsburg, in small brown crystals and crystalline granular masses.

The analyses of zincblende from Pennsylvania gave the following results:

a. Brown blende from Phœnixville, by J. L. Smith (Am. J. Sc. [2] XX, 250). *b.* Brown blende from the Ecton Mine, by Harry W. Jayne (Laboratory Univ. of Penna.). *c.* Grayish white crypto-crystalline blende from Friedensville, by the same:

	a.	*b.*	*c.*
Quartz	—	0.80	0.75
Sulphur	33.82	33.45	33.13
Zinc	64.39	65.04	66.72
Cadmium	0.98	—	—
Copper	0.32	—	—
Lead	0.78	—	—
Iron	—	2.17	0.49
	100.29	101.46	101.09

15. *Chalcocite* (*Dana*).

Chalcocite or copperglance crystallizes in orthorhombic crystals; more commonly found in granular, compact or impalpable masses. H = 2.5; sp. gr. = 5.5-5.8. Color iron-black, inclining to lead gray. In its pure state sub-sulphide of copper, Cu_2S, containing copper 79.8, sulphur 20.2.

It is a rare mineral in Pennsylvania and occurs at no locality in sufficient quantity to be valuable as an ore.

It has been found with bornite, &c., in the upper Devonian rocks (Catskill group) of Lycoming, Sullivan, etc., counties; in the neighborhood of Muncy some excavations have been made in the hope of finding it in quantity.

According to a private communication from Prof. W. Th. Roepper, of Bethlehem, chalcocite, occasionally in small crystals, is frequently met with on the line of junction between the South Mountain rocks and the triassic sandstone, which latter is often colored green by malachite, resulting from the oxidation of the chalcocite.

Small quantities have been found at the Wheatley Mine near Phœnixville, and very rarely at the Wood's Mine, Lancaster county, associated with chrome ore.

16. *Millerite* (*Haidinger*).

It crystallizes in the hexagonal system—in combinations of several rhombohedra and prisms—with perfect rhombohedral cleavage. Rarely in distinct crystals, usually in groups of capillary crystals, in tufts, semi-globular and radiated. H = 3.5; sp. gr. =4.6-5.6. Color brass yellow, inclining to bronze yellow, often tarnished. Its composition is: Sulphide of nickel = NiS. Nickel 64.9. Sulphur 35.1.

This beautiful mineral occurs rarely in yellow acicular crystals associated with quartz in the argillaceous siderite of the Carboniferous strata at Scranton (Roepper priv. com.).

In small grains and nodules imbedded in kæmmererite, it is rarely found at Texas, Lancaster county (W. P. Blake in Am. J. of Sc. [2] XIII, 116).

The most interesting locality, and one of the best in the world, is the Gap Mine, Lancaster county. Here it occurs in columnar coatings and mamillary and semi-globular radiated masses of a pale brass yellow color and metallic lustre; sometimes the millerite is partly altered into chalcocite and assumes a black color without lustre. It contains, according to my analyses (Am. Journ. Sc. [2] XXXIII, 195):

		Millerite.	Partly altered into Chalcocite.
Sulphur	=	35.14	33.60
Nickel		63.08 }	59.96
Cobalt		0.58 }	
Iron		0.40	1.32
Copper		0.87	4.63
Gangue		0.28	0.54
		100.35	100.05

Millerite is a very valuable nickel ore, which at the Gap Mine has occasionally been met with in considerable quantities.

17. *Pyrrhotite* (*Dana*).

Crystallizes rarely in hexagonal plates with basal cleavage. Mostly massive, laminated and granular. H 3.5–4.5 ; sp. gr. = 4.4–4.6. Color between brass yellow and copper red, readily tarnishing pinchbeck brown; lustre metallic—magnetic. Mostly: Fe_7S_8 = Iron 60.5, Sulphur 39.5. Frequently part of the iron replaced by nickel and cobalt.

The ordinary pyrrhotite has been found in small quantities at several localities in Pennsylvania; in crystalline plates—associated with orthoclase at Quarryville, Lancaster county; massive in hornblende rock at Morris Hill; also at the Soapstone Quarries on the Schuylkill, near Lafayette, Montgomery county; with scapolite, etc., at Geo. Van Arsdalen's Limestone Quarries, near Feisterville, Bucks county.

A slightly niccoliferous variety has been found in Alsace township, six miles east of Reading.

The most important occurrence of niccoliferous pyrrhotite, is that of the Gap Mine, Lancaster county.

This mine was discovered in 1732, and was worked to some extent for copper, but was abandoned on account of the small quantity of copper present in the ores. About 25 years ago, it was discovered that the ore of the Gap Mine, which forms a very large vein, contained a considerable quantity of nickel. It has since been worked principally for the latter metal.

Mr. Joseph Wharton kindly furnished me with the following statement:

"The quantity of nickel ore taken from the Gap Mine is as follows, reckoning the ore all as nickel ore, though in a small part of it copper predominated, viz.:

"1. By Gap Mining Co., 1853 to 1860, . 38,000,000 lbs.
"2. By myself in 1863, ... 4,500,000 "
"3. By myself from Jan. 1, 1864 to Dec. 1, 1874, 92,153,364 "

"Items 1 and 2 are not strictly accurate, as no accounts can be found showing the Gap Co.'s work, and my operations in 1863 were tentative and irregular, but they are approximately correct. Item 3 is correct."

Numerous analyses of the Gap Mine ores which I have made, gave on an average from 2–2.9 per cent. of nickel and cobalt; containing about 4 per cent. of cobalt.

Some richer varieties have been analyzed by Prof. Martin H. Boyé (Am. Journ. of Sc. [2] XIII, 219), and Prof. Rammelsberg (Mineralchemie, 113):

		Boyé.	Rammelsberg.
Sulphu	=	24.84	(38.59)
Iron	=	41.34	55.82
Nickel (and Cobalt)	=	4.55	5.59
Copper	=	1.30	
Lead	=	0.27	
Silica and insoluble silicates	=	25.46	
Alumina	=	1.07	
		99.46	100.00

I have already under Pentlandite, referred to a non-magnetic nickel-iron-sulphide found in the Gap Mine ores.

18. *Greenockite* (*Jameson*).

This very rare mineral, sulphide of cadmium, has been observed in Pennsylvania only in its amorphous variety. It was discovered many years ago by Prof. W. Th. Roepper, of Bethlehem, who found it in the form of yellow, greenish yellow and orange yellow earthy incrustations upon zincblende, smithsonite and limestone, at Friedensville; also in a clay-like material partly resulting from the decomposition of a pyritiferous rock.

A specimen of this clay analyzed by Prof. Roepper, contained 5 per cent. of cadmium, equal to 6.4 per cent. of greenockite.

Formerly cadmium was separated from the zinc at the Bethlehem works.

19. *Pyrite (Haidinger) or Iron Pyrites.*

Pyrite or bisulphide of iron is one of the commonest minerals in Pennsylvania.

It crystallizes in forms belonging to the isometric system, usually in cubes, the pentagonal dodecahedron (pyritohedron,) and octahedron, and frequently in combinations of these forms; also reniform, globular, stalactitic with crystalline surface; radiated, granular and amorphous. H = 6—6.5; sp. gr. = 4.8—5.2. Color pale brass yellow. Lustre metallic. It is bisulphide of iron, FeS_2 containing: iron 46.7, sulphur 53.3. Small quantities of other metals are frequently found in pyrite, such as gold, silver, thallium, cobalt, nickel, copper, &c.

It is usually found in small brass yellow cubes in many of the rock-masses, especially in slates; also in most of the coal mines, both anthracite and bituminous, and very frequently in the numerous iron mines in the State.

Although of such a general occurrence, there is nowhere a sufficiently extensive and massive deposit of it to be useful for the manufacture of sulphur and sulphuric acid.

It would be tiresome to enumerate all the localities where it has been observed, I shall confine myself, therefore, to a few where interesting and charactistic varieties have been found.

Cubical crystals sometimes of great brilliancy, which have frequently been mistaken for gold or gold ore, often partly oxydized and of a brown color, are the most common form and very abundant in the neighborhood of the city of Lancaster; also near the Valley Store in Tredyffrin and in E. Whiteland townships, and in clusters near Oxford, Chester county. Cubic crystals are also found at the beryl locality in Newlin and East Bradford townships, Chester county, and at Chester, Delaware county.

Very beautiful brilliant octahedral crystals, from small size to about an inch in diameter, are abundant at several of the iron mines near Knauertown, Warwick township, Chester county;

in octahedral crystals, from ½ to 2 inches diameter, it has been found at Van Arsdale's Quarry, Bucks county.

Cubes and cubo-octahedra have been met with at the Poorhouse Quarry and John Bailey's Quarry, East Marlboro', Chester county; also in various forms at Perkiomen Mine, Montgomery county; near Phœnixville, in the Tunnel as well as in the Silverlead Mines; and Jones Mine in Berks county; good cubical crystals with octahedral planes are occasionally met with at the Gap Mine, Lancaster county; small cubo-octahedra occur sometimes with the zincblende of Friedensville in Lehigh county; cubes and also pentagonal dodecahedra occur in the steatite of Chestnut Hill, near Easton.

Very interesting and beautiful crystals are found in the talcose slates on M. Boice's farm, one mile northwest of Texas, Lancaster county; all the crystals from this locality have a brownish tarnish, and are combinations of cubes with the octahedron and the pentagonal dodecahedron.

Very complicated and much distorted forms have been observed at Cornwall, Lebanon county.

Peculiar incrustations and stalactitic prolongations of pyrite are found in the mines of the Westmoreland Coal Company, Westmoreland county.

In most of these localities the pyrite is, if not oxydized, nothing but the pure bisulphide of iron.

I will now mention a few varieties containing small quantities of other metals.

Professor W. T. Roepper has made the interesting observation that the octahedral crystals or those in which the octahedral planes are predominating, invariably contain cobalt.

A cobaltiferous variety from Cornwall, Lebanon county, afforded J. M Blake 2 per cent. of cobalt (Dana's Min. 68).

A niccoliferous pyrite from the Gap Mine, which occurred there in 1861, in granular porous masses without distinct form, which, however, did not show any admixture of millerite, contained, according to my analyses, 4.30 and 4.47 per cent. of nickel;

A cupriferous variety from Cornwall, Lebanon county, which

readily tarnishes and assumes a steel-blue color, contains, according to J. C. Booth (Dana Min. 63):

Sulphur,	=	53.37	per cent.
Iron,	=	44.47	"
Copper,	=	2.39	"
		100.23	

The flue-dust of the iron furnaces at Bethlehem contains, according to Prof. Roepper, a considerable quantity of thallium, which he attributes to thalliferous pyrite contained in the Pennsylvania anthracite used.

20. *Chalcopyrite (Henkel), or Copper Pyrites.*

It crystallises in the tetragonal system; good crystals are not common; they are mostly combinations of several tetrahedra. Easily distinguished from pyrite by its greater softness. H = 3.5 –4; sp. g. = 4.2; color brass-yellow, often tarnished and iridescent. Lustre metallic. Often massive. It is a combination of sub-sulphide of copper with sesqui-sulphide of iron = $Cu_2S + Fe_2S_3$, containing, when pure, copper 34.6, iron 30.5, sulphur 34.9.

Small quantities of this valuable copper ore have been obtained from some of the mines in Triassic sandstone and shale, in Bucks county about three miles west of New Hope; at the Perkiomen Mine and Jug Hollow Mine, Montgomery county, and the Morris Mine, near Phœnixville; it is also associated with the silver lead ores of the Chester county mines, especially the Wheatley Mine, where fine tetrahedral and octahedral crystals were obtained.

It occurs in small seams and irregular masses disseminated through magnetic iron ore at some of the Mines near Knauertown, in Chester county; the Jones Mine, near Morgantown, and Fritz' Island Mine, near Reading, in Berks county; and the Cornwall Mines in Lebanon county.

Massive or in minute quantities disseminated through the nickel ores of the Gap Mine, and as an associate of the molybdenite at Chester, Delaware county, also in the gneiss at Frankford and Wissahickon near Philadelphia and the Soapstone Quarries on the Schuylkill, near Lafayette, Montgomery county.

Very fine crystals, large tetrahedra, from ½ to ¾ inches in size, but often tarnished, and coated with malachite, occur in the magnetic mines already mentioned, especially at the Jones' Mine, also near Knauertown, and at Cornwall.

The chalcopyrite from the Wheatley Mine has been analysed by J. L. Smith, in his Minerals of the Wheatley Mine (Am. Journ. Sc. [2] XX. 242-253). It contains:

Copper	=	32.85
Iron	=	29.93
Lead	=	0.35
Sulphur	=	36.10
		99.23

21. *Gersdorffite* (*Loewe*).

It crystallizes in the isometric system, and consists of sulphur, arsenic and nickel, part of the latter being frequently replaced by cobalt.

I have observed (Am. Jour. of Sc., [2] XXVIII, 248), this very interesting mineral in minute quantities as an associate of anglesite, galenite, zincblende and quartz, at the Wheatley Mine, near Phœnixville. The only specimen which is known consists of an incrustation of minute crystals, which are combinations of cubes, octahedra and rarely the pyritoid. They have a grayish white color, and have been found to be principally a sulpharsenide of nickel, containing, however, a considerable quantity of cobalt.

22. *Marcasite* (*Haidinger*).

The rhombic bisulphide of iron is said to occur at the Gap Mine and according to Prof. W. T. Roepper, in large masses in many of the iron mines of the Lehigh region.

I have never seen it from any locality in Pennsylvania, but the fact that many varieties of sulphide of iron found with the limonite oxydize very readily and change to copperas appears to confirm Prof. Roepper's observation.

23. *Leucopyrite* ? (*Shepard*).

In Dana's mineralogy it is mentioned that a crystal of arsenical iron, of two or three ounces in weight, had been found in Bedford county. This is Prof. Shepard's "meteoric iron."

In a letter from Prof. Shepard, of December 23d, 1874, he says: "The Bedford county ore I have long since withdrawn from my list of meteorites, and think it belongs to *leucopyrite*, unless it is a new species."

If we compare this with his former statement that he had *not* been able to confirm the presence of arsenic or any other metal, we cannot conceive how he now can place it with leucopyrite, a mineral which contains in its pure state 72.8 per cent. of arsenic. I have no doubt that my conclusion, that the Bedford county iron is neither a meteorite, nor leucopyrite, nor a new mineral, but a piece of "spiegeleisen" is the correct one.

24. *Arsenopyrite* (*Glocker*), *or Mispickel.* ?

This is another of the doubtful Pennsylvania minerals.

Isaac Lea in "an account of minerals at the present known to exist in the vicinity of Philadelphia" (Jour. Ac. Nat. Sc., Phil., I, 462, Dec., 1818), states that a piece of this mineral of nearly two pounds weight had been given to him by a person at Perkiomen, who informed him that it was from the neighborbood.

It had a yellowish white color, uneven fracture, and emitted, when subjected to the blowpipe, white vapors of arsenious acid and a strong alliaceous odor.

25. *Covellite* (*Freiesleben*).

This beautiful mineral crystallizes rarely in hexagonal plates, generally it is massive, crystalline and in thin coatings on other minerals. It has a rich indigo and steel-blue color, and contains in its pure state, 66.5 per cent. of copper and 33.5 per cent. of sulphur. It occurs in no place in sufficient quantity to be useful as an ore.

It is found at the Perkiomen Mine, very rarely in perfect, though microscopic hexagonal plates, usually in crystalline masses of a deep blue color; also in small quantities at the lead and copper mines, near Phœnixville, and as a thin coating of magnetite at the Cornwall Mine, Lebanon county.

C. *Sulpharsenides, Sulphantimonides, etc.*

26. *Tetrahedrite* (*Haidinger*).

Prof. Samuel P. Sadtler stated to me that small quantities of *tetrahedrite* used to be found in the epidote rock, just south of Gettysburg, associated with native copper and malachite, but that the locality is now exhausted.

It has not been investigated to which of the varieties it belongs.

27. *Tennantite* (*Phillips*).

In a notice of two minerals from Lancaster, Pa., Zinc Mines, a few miles northwest of Lancaster City, by W. J. Taylor (Am. Jour. of Sc. [2] XX, 412), the occurrence of tennantite is described.

It is massive, steel gray, yielding a dark reddish gray powder. A qualitative analysis showed the presence of sulphur, arsenic, copper, iron and zinc.

CHAPTER III.

Compounds of Chlorine, Bromine, Iodine.

28. *Halite* (*Dana*) *or Common Salt.*

No rock salt deposits have yet been found and developed in Pennsylvania, but in many localities in the western part of the State a considerable number of salt wells have been found, from which, to a limited extent, salt has been manufactured. Many brine springs have also been found when boring for oil.

The information which I have been able to obtain with regard to the composition of these salt-water springs is very limited.

Dr. Edward Stieren has published "Observations on the Salt waters of the Alleghany and Keskiminetas Valleys" (Am. Journ. Sc. [2] XXXIV, 46–57), and has given an analysis of the water from the salina of Mr. Peterson, in the vicinity of Tarentum, in which he found the following number of

		Grains in one pound of 7680 grains,		or in 1000 parts.
Chloride of sodium	=	253.519827	—	33.010394
" potassium	=	0.312945	—	0.040748
" ammonium	=	0.051571	—	0.006715
" barium	=	0.047055	—	0.006127
" strontium	=	0.736949	—	0.095970
" calcium	=	65.944942	—	8.586581
" magnesium	=	16.654496	—	2.168541
Bromide of magnesium	=	0.884175	—	0.115127
Iodide of calcium	=	0.618055	—	0.080476
Carbonate of baryta	=	0.029084	—	0.003787
" strontia	=	0.455685	—	0.059334
" lime	=	21.246190	—	2.766431
" magnesia	=	10.582281	—	1.377901
" iron	=	0.295349	—	0.038457
" manganese	=	traces	—	traces
Silicate of alumina	=	0.313881	—	0.040870
Free silicic acid	=	0.787515	—	0.102541
		372.480000	—	48.500000
Carbonic acid, loosely bound as bicarbonate	=	14.998187	—	1.952889
" " really free	=	0.060572	—	0.007887
		387.538759	—	50.460776

An analysis which I had made in my laboratory about fifteen years ago, by Mr. Geo. J. Poepplein, to determine the principal constituents of salt water from the neighborhood of the Conemaugh, near Saltzburg, gave in 1000 parts:

Chloride of sodium	=	71.320
" magnesium	=	3.986
" calcium	=	15.726
Bicarbonate of calcium	=	0.C05
" iron	=	0.078
		91.115

The other constituents were not determined, but their nature appears from an analysis of the mother-liquor from the manufacture of salt (or "Bittern") on the Kiskiminetas River near Freeport, Armstrong county, by M. H. Boyé (read 1848, at the Philadelphia meeting of the Am. Assoc. for Advancement of Science), who found in 100 parts:

Chloride of potassium	=	0.128
" sodium	=	0.877
" calcium	=	24.640
Chloride, bromide and iodide of magnesium	=	10.146
		35.791

The latter contained:	Magnesium	=	2.5750
	Chlorinə	=	6.8660
	Bromine	=	0.7010
	Iodine	=	0.0035
			10.1455

A partial analysis which I have made of a salt water, found in boring for oil near Alba, Bradford County, gave in 1000 parts:

Chloride of sodium	=	80.3770
" iron	=	trace
" calcium	=	13.3651
" magnesium	=	1.9075
Bicarbonate of iron	=	0.0986
" lime	=	0.0374
Silicic acid	=	0.0530
Bromine, &c., not determined	=	——
		95.8386

The quantity of bromides which many of the Pennsylvania salt waters contain is very remarkable, and larger than in any other brines in the world.

A very large amount of the bromine of commerce, so extensively used at present in photography and for the manufacture of bromide of potassium, is extracted from the mother-liquor of the salt works, and a large portion exported to Europe.

I will mention here the analysis of a very extraordinary mineral water, which has been obtained in boring for oil in Elk county.

It has been called the "*East Clarion Spring Water*," and contains, according to an analysis which I have made, in one gallon of 231 cubic inches, as follows:

Chloride of potassium	=	0.89971	grains.
" sodium	=	336.80275	"
" lithium	=	trace	
" barium	=	1.72573	"
" strontium	=	0.06260	"
" calcium	=	51.85625	"
" magnesium	=	15.34206	"
Nitrate of ammonia	=	0.19172	"
" magnesia	=	0.13623	"
Phosphate of lime	=	trace	
Bicarbonate of baryta	=	0.12791	"
" strontia	=	0.00487	"
" lime	=	9.79502	"
" magnesia	=	0.57955	"
" iron	=	0.72444	"
Silicic acid	=	0.69523	"
		418.94407	"

This mineral water contains the largest quantity of chloride of barium ever observed in any, and may become of great importance after its medicinal properties have been more fully investigated.

Although not a chloride of sodium water, as it contains less than a grain of this substance in a gallon, I will here add the analysis which I have made of the "Gettysburg Katalysine" water, about which so many contradictory statements have been circulated.

Particular attention was paid to the determination of lithia and such constituents, which are present in the most minute quantities, and for this purpose over 24 gallons of it were evaporated.

I found in one gallon of 231 cubic inches, as follows:

Sulphate of baryta	=	trace	
" strontia	=	0.00427	grains.
" lime	=	0.83145	"
" magnesia	=	6.77940	"
" potash	=	0.20836	"
" soda	=	2.46776	"
Chloride of sodium	=	0.65790	"
" lithium	=	trace	
Bicarbonate of soda	=	0.70457	"
" lime	=	16.40815	"
" magnesia	=	0.54260	"
" iron	=	0.03585	"
" manganese	=	0.00669	"
" nickel	=	trace	
" cobalt	=	trace	
" copper	=	0.00050	"
Borate of magnesia	=	0.03492	"
Phosphate of lime	=	0.00679	"
Fluoride of calcium	=	0.00954	"
Alumina	=	0.00380	"
Silicic acid	=	2.03078	"
Organic matter with traces of nitric acid, &c.	=	0.70870	"
Impurities suspended in the water, like clay, &c.	=	1.10069	"
		32.54272	"

As an analysis of the Gettysburg Katalysine water is published, which purports to contain a far larger quantity of constituents, other examinations of *genuine* water have since been made by me, which corroborated my previous results.

There are numerous other mineral springs in the State, but I have not been able to find any recent analyses of them.

29. *Sal Ammoniac.*

Chloride of ammonium, or as it is generally called, sal ammoniac, is found in minute dodecahedral crystals and crystalline crusts and fibrous masses of a white or yellowish-white color at the Burning Mine, at Summit HIll, Carbon county. It is frequently colored yellow by ferric chloride.

CHAPTER IV.

FLUORINE COMPOUNDS.

The only mineral belonging to this class, which occurs in this State is:

30. *Fluorite* (*Napione*).

Fluorite or fluorspar crystallizes in the isometric system and, if not in the pulverulent or crypto-crystalline massive varieties, shows a very eminent octahedral cleavage. Its hardness = 4; sp gr. = 3.18. Its colors are variable, from colorless through almost all shades of yellow, green, blue or red. Lustre vitreous. In its pure state consists of fluoride of calcium $CaFl_2$. = Calcium 51.3, Fluorine 48.3.

It is a rare mineral in Pennsylvania.

Minute, but beautiful, colorless or pale yellowish white cubes and combinations of the cube and tetra-hexahedron, frequently inclosed in calcite or upon the apex of scalenohedral calcite crystals, also crystalline incrustations of fluorite have been found at the Wheatley Mine, near Phœnixville.

It has been analyzed by J. L. Smith (*l. c.*), who found:

Sp. gr.	=	3.15
Fluorine	=	48.29
Calcium	=	50.81
Phosphate of lime	=	trace.
		99.10

Purple colored fluorite is found in small cubes and crystalline masses in Edwards old limestone quarry, in Newlin township, Chester county.

A similar variety occurs at the quarries above Delaware Water Gap; small particles of crystalline violet fluorite have been observed in the limestone quarries at Ironton, Lehigh county.

It is rarely met with in purplish crystalline masses in the gneissic rocks at Frankford, and the Falls of the Schuylkill, near Philadelphia.

It is mentioned in Dana's mineralogy, as having been observed in Lebanon county.

CHAPTER V.

OXYGEN COMPOUNDS.

A. Binary Oxygen Compounds, or Oxides.

I. *Oxides of Elements of Series I.*

a. Anhydrous.

31. *Cuprite* (*Haidinger*).

It crystallizes in the isometric system, usually in octahedra, cubes or dodecahedra, or combinations of these forms; the crystals are frequently so much distorted and elongated that they become capillary or needle-shaped. The latter variety has been called chalcotrichite. It also occurs massive. H. = 3.5–4; sp. gr. = 5.85–6.15. It has a red color, sometimes inclining to brown, frequently cochineal or crimson red and sub-metallic and adamantine lustre. In its pure state it is sub-oxide of copper, Cu_2O, containing 88.80 per cent. of copper and 11.20 per cent. of oxygen.

It has been observed at several localities in the State, but nowhere in sufficient quantity to be valuable as an ore, although it greatly helps to enrich the poorer copper ores, especially at Cornwall.

It occurs on magnetite at Cornwall, Lebanon county, in crystalline finely granular masses and beautiful octahedral crystals, sometimes showing cubical and dodecahedral faces and in fine crimson red capillary crytals (chalcotrichite), also in very thin coatings upon native copper, giving the latter a dull, somewhat purplish appearance.

It was occasionally met with at the Perkiomen Mine, especially the variety chalcotrichite in very fine acicular crystals, sometimes one inch in length.

The massive variety associated with the native copper is found in Adams county, in epidotic rocks, and in small quantities at Chestnut Hill, near Easton.

I have found an old analysis, by Henry Seybert (Jour. Ac.

Nat. Sc., Phila., II, 2, 142), of a variety from Cornwall, which evidently shows a large admixture of magnetic iron and clay, but which had been formerly considered a *distinct* mineral, "a ferruginous oxydulated copper ore," in which he found:

Water	=	6.98
Clay (Argile)	=	3.80
Deutoxide of iron	=	42.16
Protoxide of copper	=	43.88
		96.82

32. *Water.*

I omit any remarks about this substance, especially as I have given the analyses of waters more or less charged with mineral ingredients, under the heading of *Halite.*

33. *Melaconite* (*Dana*).

The black oxide of copper CuO, when found as a mineral has been called melaconite; if pure, it contains copper 79.85. oxygen 20.15.

The pulverulent and massive variety occurs in this State at the Perkiomen mine in Montgomery county, and the silver-lead mines, near Phœnixville, in Chester county.

It has not been analyzed.

34. *Corundum.*

Crystallizes in the rhombohedral system; many combinations have been observed at foreign localities.

The corundum, which occurs in Pennsylvania, when crystallized is usually found in acute bi-pyramidal crystals of a rough or uneven surface, either isolated or in clusters and of gray, grayish white, grayish brown, bronze or blue colors. It shows both the rhomhedral cleavage at an angle of 86° 4′, and that parallel to the base; it is also found granular and massive; H = 9; sp. gr. = 3.9–4.16. In its pure state it is pure alumina, but generally contains minute quantities of other substances.

The collections of Mr. W. W. Jefferis, of West Chester; Mr. Lewis Palmer, of Media; Dr. Cardesa, of Claymont, Del.; Col. Joseph Willcox and Dr. Isaac Lea, are rich in highly interesting varieties.

At the Black Horse, near Media, in Middletown township, Delaware county, it has been found in a feldspathic rock (probably a variety of massive oligoclase), in slender bi-pyramidal grayish crystals; some show on the cleavage plane a bronze lustre and are frequently asteriated; near the same locality it is often found loose in the soil, sometimes in large crystals from four to six inches in length.

At Mineral Hill, near Media, a feldspathic rock is penetrated by slender crystals of corundum, more or less completely altered into fibrolite.

On the north slope of Mineral Hill, it has been found in large brown crystals, to a great extent changed into massive margarite.

Near Village Green, in Aston township, Delaware county, it occurs in large crystals, of a brownish color, often with a bronze lustre, and completely surrounded by scaly margarite, into which it is partly changed.

A small mass of grayish white and blue corundum which shows perfect cleavage has been found in the neighborhood of the chrome mines, near Texas, Lancaster county.

In Chester county it has been found in crystals associated with a feldspathic mineral (? albite) near Fremont P. O., West Nottingham township, and in the same manner about two miles south of Oxford.

Near Unionville, in Newlin township, are several localities of interest and importance.

About one mile and a-half northeast from this place it occurs in gray crystals in white granular albite, associated with tourmaline, euphyllite and several other minerals.

Loose crystals of 3 to 4 inches in length, also clusters of crystals of a white, grayish-white, gray and blue color, often showing a perfect cleavage, have frequently been found in the soil. Many of them are surrounded by a coating of soda margarite and damourite, into which it has been partly converted.

Loose masses of brownish-gray granular variety, sometimes of several tons in weight, were occasionally met with at the same locality.

About two years ago the deposit from which the latter evidently have come was discovered. It was at the time of my visit exposed to a length of about 30 feet, a thickness of from five to ten feet with a depth of about fifteen feet.

This is a very important discovery, and the appearance of the deposit or bed leads to the conclusion that it is the nucleus of unaltered corundum in a mass of chlorite slates and other rocks which have resulted from its alteration (see my investigations on Corundum, its alterations and associated Minerals, —Proc. Am. Phil. Soc. XIII, 361–407).

From present indications this corundum mine appears to be of great value, and promises to furnish large quantities of it of a very superior quality.

In the use of corundum as an abrasive material it is of great moment whether or not it is easily cleavable; if the minute particles lie flat upon the material which is to be cut or polished, they have not near as much effect as those of a corundum which, with the same hardness, cleaves less readily, and always presents a sharp or cutting edge. This latter is the case with the coarse granular variety from Unionville more than with that from any other part of the world.

It is now worked up at Kennett Square into the different sizes mostly required by the trade and on account of its great superiority over *emery* must soon drive the latter from this market.

At none of the other localities above enumerated has corundum been found in sufficient quantity to be useful in the arts; some has been obtained from the neighborhood of Media and worked up into corundum wheels, &c.

35. *Hematite* (*Hausmann*).

Like corundum, hematite crystallizes in the hexagonal system, usually in combinations of several rhombohedra, scalenohedra and the basal plane, with distinct cleavages parallel to the rhombohedron of 86° 10′ and the base. Also columnar, granular, botryoidal and stalactitic; lamellar, foliated, micaceous to fine scaly, fibrous and compact. Lustre from metallic to dull. H = 5.5–6.5; sp. gr. = 4.5–5.3. Color dark steel-gray to iron-black, in thin particles blood red; the fibrous and earthy varieties dull red. In its pure state, ferric oxide = Fe_2O_3, containing 70 iron and 30 oxygen.

A few very beautiful crystals in possession of Mr. Samuel Tyson have been found associated with ripidolite on chromite, at Wood's Mine, Lancaster county.

A lamellar iron-black variety is very common in many localities in the southeastern part of the State in Montgomery, Delaware, Chester, Lancaster, York and Adams counties; it is usually associated with quartz and often mistaken for menaccanite, but the titanic acid which it contains is usually very small. Dr. Wm. P. Headden, Assistant in the Laboratory of the University of Pennsylvania, has determined the amount of titanic acid in that from Edgehill, Montgomery county, and found 0.93 per cent. of it.

It occurs nowhere in the State in sufficient quantities to be valuable as an iron ore.

Foliated micaceous hematite is occasionally met with in the magnetic mines at Cornwall, Lebanon county, the Steele Mine, near St. Mary's, and the mines near Knauertown, in Warwick township, Chester county, and in York county.

Micaceous hematite of an iron-black and of a red color, occurs at New Salem, seven miles from Gettysburg, also north of Gettysburg, where quite extensive beds are now opened from which ores are sent to Gettysburg for shipment.

Extensive deposits of micaceous hematite in which this mineral replaces mica in a micaceous schist, a sort of "itabyrite," also containing magnetite, occur in York county in the so-called Codorus region.

Red hematite has been found near Durham, in Bucks county, and in some of the iron mines of Berks county.

The scaly or micaceous variety occurs along the junction of the South Mountain rocks, and the new red sandstone; also in Schuylkill and Dauphin counties; at Perkiomen, Montgomery county, and the Silverlead Mines near Phœnixville.

The most important occurrence of red hematite in Pennsylvania is that of the argillaceous variety, the so-called red fossil ore in several formations.

The fossil ore of the Clinton group is found on both flanks of the Montour ridge in Montour and Northumberland counties; the Buffalo Valley, Union county; the south flank of Jack's Mountain, and north flank of Shade Mountain, in Snyder and

Mifflin counties; the great Aughwick Valley, Huntingdon county; Tuscarora Mountain, Juniata county; the north flanks of Muncy, Bald Eagle, Lock, Dunning and Wills Mountains, extending through Lycoming, Centre, Blair and Bedford counties, and the south flank of Tussy Mountain, in Huntingdon, Blair and Bedford counties, and the north flank of Stone Mountain in Huntingdon county.

The Medina fossil rock ore underlying the last about one thousand feet, usually accompanies the same and occurs independently at the Schuylkill Water Gap.

The fossil ore of the Catskill group has been developed at Larey's Creek, in Lycoming county, and Deeler's Creek, in Somerset and Wayne counties.

All the fossil ores show frequently, especially near their outcrop, an alteration into the hydrated ferric oxide or limonite.

Hematite is occasionally found in the neighborhood of Lancaster and elsewhere, in cubical crystals, which are pseudomorphs after pyrite.

Martite (*Breithaupt*).

This name was given to an *octahedral* variety of hematite, which, by some mineralogists, is considered a good species, and the dimorphous form of ferric oxide, but it is more probably a pseudomorph after magnetite.

It is not magnetic and yields a red powder.

Prof. W. Th. Roepper has observed that it is frequently associated with zircon, allanite, etc., at several of the South Mountain localities near Bethlehem.

36. *Menaccanite* (*McGregor*).

It crystallizes in the hexagonal system, usually in combinations of the hexagonal prism with several rhombohedra and scalenohedra; its cleavage is basal. Crystals are often tabular or in thin plates and laminae; occurs also in loose grains or sand. H = 5–6; sp. gr. = 4.5–5. Iron-black with sub-metallic lustre. Yields a black or brownish red powder. The general formula for menaccanite is $(FeTi)_2O_3$, or, in other words, a ferric oxide, in which a *variable* quantity of iron is replaced by titanium, the latter varying from 3.5 to 59 per cent.

A very beautiful crystal, a combination of a hexagonal prism and rhombohedron and of an iron-black color has been observed in a mass of gneiss in West Philadelphia, near the Columbia Bridge, opposite the Fairmount Water-works. It is about 1½ inches in diameter, and nearly a half inch in thickness.

A large crystal, showing distinctly the same planes from Dutton's mills, near Upland, Delaware county, is in Dr. Cardesa's cabinet; it is about four inches long, one and a-half inches broad and three-quarters of an inch thick.

A very fine crystal and smaller crystalline plates, from Marple, Delaware county, are in Col Willcox's cabinet.

With muscovite it is found in Elk township, Chester county, with ripidolite, at Patterson's, in Newlin township, Chester county.

Inferior specimens occur near Megargee's paper mill, and Heft's mill, near Philadelphia, and in Middletown township, near Media, Delaware county.

In Chester county it has been met with in Westtown township, and on Joseph H. Brinton's farm, Thornbury township, and in blue quartz with zircon, near the paper mill on the Brandywine, and loose in the soil about one and a-half miles from the allanite locality in East Bradford township.

It is found in the sands of the Delaware and Schuylkill Rivers, and probably in those of many of the other streams. Menaccanite is not found in Pennsylvania as a massive iron ore.

37. *Spinel.*

It crystallizes in the isometic system, in octahedral crystals frequently modified; also granular, massive. H = 8; sp. gr. 3.5–4.9. Colors red, blue, green, yellow, brown, black. The typical spinel is a combination of alumina with magnesia $= MgO,Al_2O_3$; in the various varieties, the magnesia is frequently in part replaced by ferrous and sometimes manganous oxide, rarely by lime, the alumina by ferric oxide; pleonaste and hercynite are the names of black and greenish-black spinel varieties.

Dr. George Smith (History of Delaware county, Philadelphia, 1862, page 415), mentions spinel as occurring in minute octahedral crystals at Blue Hill, Delaware county.

I have observed it at the corundum locality one and a-half miles northeast of Unionville, associated with granular grayish white talc, and greenish actinolite, and intimately mixed with chlorite and particles of corundum, from the alteration of which it evidently results.

It has a black color, vitreous lustre and a hardness superior to that of quartz.

An analysis of material selected as carefully as possible, but still containing a considerable admixture of chlorite and corundum, made by Dr. G. A. Koenig (my paper on Corundum, etc., *l. c.*), gave the following results:

Alumina	=	54.61
Ferric oxide	=	4.10
Ferrous "	=	10.67
Magnesia	=	13.83
Silicic acid	=	1.26
Corundum	=	16.24
		100.71

This spinel is the variety "pleonaste" probably with an admixture of "hercynite."

The occurrence of spinel associated with allanite in the South Mountains, five miles east of Bethlehem, has been observed by Prof. W. Th. Roepper (priv. com.).

38. *Magnetite (Haidinger).*

Magnetite crystallizes in the isometric system; the usual form is the octahedron, often in perfect crystals and hemitropes dodecahedral crystals are also of frequent occurrence, generally with striation of the planes parallel to the longer diagonal; also granular, massive; cleavage octahedral, sometimes very distinct. H = 5.5–6; sp. gr. = 5–5.2; color iron-black, in very fine particles as occurring in muscovite, various shades of brown; powder black; fracture uneven, sometimes subconchoidal. Magnetic and sometimes polaric (loadstone).

Pure magnetite consists of a combination of ferric and ferrous oxides = FeO, Fe_2O_3, containing 72.4 per cent. of iron and 27.6 per cent. of oxygen. In some varieties part of the ferrous oxide is replaced by magnesia, in others a part of the iron in the ferric oxide by titanium as in menaccanite.

Pennsylvania possesses many very important deposits of magnetite, stretching from the Durham Hills, south of Easton, along the South Mountain range, across the State. Very important mines of this ore in Berks county, are near Reading, at Topton, Boyertown, Mt. Penn, Fritz's Island and others; the Jones Mine, near Morgantown; in Chester county, the magnetite mines in Warwick township, the Hopewell Mines, the mines near Knauertown and St. Mary's; in York and Adams counties; and most important of all, the Cornwall Mines in Lebanon county.

It would be tedious to enumerate all localities where this mineral has been found. I will mention therefore only a few, where varieties have been observed which are interesting on account of their crystallization or the peculiarity of their occurrence.

Very brilliant octahedral crystals occur at the Hopewell and Steele's Mine in Warwick, also two miles north of Unionville in Newlin township, Chester county; large and perfect octahedral crystals, simple and hemitropes are found in the chlorite slate near Texas, Lancaster county; near Easton; at Chestnut Hill, near Philadelphia, and in the soapstone quarries near Lafayette, Montgomery county; at Frankford, in the gneiss, it is rarely met with in octahedral crystals, but frequently in irregular cleavable patches.

Very good dodecahedral crystals occur at the Jones Mine, near Morgantown, Berks county, and near St. Mary's, Warwick, Chester county.

Magnetite is found in crystals and crystalline particles in the micaceous hematite schists and chloritic rocks and elsewhere in York county, also massive and in octahedral crystals at several localities near Gettysburg, Adams county.

Load-stone in fibrous masses has been found in East Goshen township, Chester county, and a granular variety at Cornwall, Lebanon county.

Reticulated magnetite is frequently met with in thin dendrites in muscovite; it is sometimes almost colorless to smoky brown and black. The best localities for this curious variety are the neighborhood of Fairville in Pennsbury, Chester county, and Middletown, Delaware county.

A peculiar variety of magnetite in foliated masses, perhaps pseudomorphs after hematite, frequently interlaminated with pyrite, has been found at Keim's Mine, near Knauertown; also at Cornwall, Lebanon County.

Detached masses of magnetite of several hundred pounds in weight are sometimes found in Brinton's serpentine quarries in Westtown township, Chester county.

Magnetic iron sand is found in all the rivers and streams of Eastern Pennsylvania.

I will mention here the occurrence of a peculiar ore which probably belongs here; it is a titaniferous magnetic iron ore, which I have observed in beds near chromite at several localities in the serpentine range.

It is iron black, massive, with uneven fracture and very lustrous; it has been found near Moro Phillips' Chrome Mine, in Marple township, Delaware county, also at several places near Texas, Lancaster county, especially at Jenkins' Mine near the Red Pit Chrome Mine, near Pleasant Grove and Soapstone Hill.

It has been best developed in Harford county, Maryland, where it has been mined and a considerable quantity of ore removed.

As there is no difference in appearance, I will give, for comparison, the analysis of an average sample of the Maryland ore, which I made a short time ago.

It contains:

Silicic acid	=	2.60
Titanic acid	=	6.60
Alumina	=	11.96
Ferric oxide	=	51.98
Chromic oxide	=	2.41
Ferrous oxide	=	18.65
Ni Co and Mn oxide	=	0.22
Magnesia	=	3.80
Water	=	1.78
		100.00

There is neither sulphur nor phosphorus present.

39. *Chromite* (*Haidinger*), *or Chromic Iron Ore.*

Isometric and crystallizing in small octahedral crystals rarely with dodecahedral planes, mostly massive, fine granular and compact. H = 5.5 ; sp. gr. = 4.5 on an average ; lustre submetallic to resinous ; color black, inclining to brown ; powder brown ; fracture uneven ; not magnetic when pure.

Throughout the whole serpentine range, which lies in the gneissic rocks of Southern Pennsylvania, crystals of chromite are disseminated through the serpentine, and after the disintegration and decomposition of the latter have given rise to the chrome sand gravel deposits. All these deposits appear to be diluvial. They are generally covered by a few inches to several feet of alluvium, which contains chrome sand in too minute a quantity to pay for its separation. Below this is a regular deposit of gravel, generally in the flats or streams and thinning out towards the hills, having a variable thickness from a few inches to four or five feet.

A few small chrome-sand gravel deposits occur in Delaware county and have furnished some very superior ore, others in the same county contain too large an admixture of garnet to be of much value.

The largest operations have been carried on in Elk, East Nottingham and West Nottingham townships, of Chester county, and at Rock Springs, and a few other branches in Lancaster county. These deposits are exhausted, or nearly so, except a few near Octarara Creek, in West Nottingham township, covering perhaps twenty acres, and containing probably 10,000 tons of chrome sand, yielding from 40 to 45 per cent. of chromic oxide ; another valuable sand chrome deposit, "Boons" occurs in Elk township, but I have not been able to obtain any particulars with regard to its probable yield.

There are probably yet 25,000 tons of chrome sand which could be made available in the deposits in Delaware, Chester and Lancaster counties.

Massive chromite, and chromite with a large admixture of serpentine, so-called birds-eye ore, forms lenticular vein-like deposits in the serpentines of Delaware, Chester and Lancaster counties. Many are only very superficial, and have produced but a few tons of marketable ore, others have been worked very extensively and for a very long time, and, together with the chrome sand from the gravel deposits, have for a very long period furnished the larger portions of the chrome ores which were required to supply the world with the raw material for the manufacture of bi-chromate of potassium and various chrome colors.

At present the state of the chrome ore business in this State is in a deplorable condition; the only mines, where work was done last summer were the Wood's Mine and Low's Mine (or Line Pit), in Lancaster county, both owned by the Tyson Mining company, of Baltimore; and the only stream, where gravel washing was carried on was the "Black Run," near the Octarara, which was worked by Mr. E. Mortimer Bye, of Wilmington, Del.

From the best and most reliable information which, in the absence of records, I could obtain, and which is certainly an approximation to the truth, the following quantities of chrome ore have been mined in Pennsylvania:

I. *Rock Chrome Ore Mines in Lancaster County.*

		TONS.
1. Low's mine or Line Pit (depth, 240 feet)............	about	700
2. Red Pit (considered the best mine, except Wood's)..		?
3. Rock Spring Mine (over 200 feet depth—much bird's-eye ore)..	about	1,000
4. Little Horse Shoe Mine.............	"	30
5. Wood's Mine (depth, 700 feet—galleries about 1000 feet long).........................		120,000
6. Carter's Mine (depth about 200 feet)...............		400

II. *Rock Chrome Mines in Chester County.*

		TONS.
7. L. Melrath's Pits, (three pits worked thirty years ago, in West Nottingham)........................	about	50
8. Scott's Mine (West Nottingham township)..........	"	3,000
9. Moro Phillips' Mine, (West Nottingham), 110 feet deep; galleries about 80 feet long................	"	250
10. Smith Hilaman's, (West Nottingham).............		?
11–13. Three old pits, in the "White Barrens," (Elk township)..	"	150
14. Amos Pugh's Mine, (East Nottingham township)..	"	20
15. Bailey's Mine (Newlin township), rock and sand ore,	"	50

III. *Rock Chrome Mines, in Delaware County.*

		TONS.
16. Black Horse Mine (Middletown township), 75 feet deep...	about	50
17. Walter Green's Mine (Middletown township).......	"	50
18. Moro Phillips' Mine (Marple township)............	"	25

IV. *Sand Chrome Ore.*

		TONS.
1. Near Rock Springs (Lancaster county),..............	about	1,000
2. Pine Barrens (West Nottingham, Chester county),	at least	1,500
3. White Barrens (East Nottingham, Chester county),	"	3,000
4. Hibbard's (near Media, Delaware county)..........	about	75
5. Fairlamb's " "	"	200
Total production....................	about	131,550

There are a great many small openings, which have produced a few up to five and twenty tons, which have not been included in the above; the total production of the chrome ores in Pennsylvania, to date, probably does not fall below 150,000 tons, of an average yield of from 40 to 45 per cent. of chromic oxide.

Some of the chrome mines which have been explored to some extent look favorable for a future supply, others are completely exhausted and the entire lenticular deposit has been removed; but no matter what the value of the mines may be, as long as the best quality of chrome ores can be brought from California or Siberia for less than the production of ours would cost, there is no prospect of reviving the mining operations for these ores.

I will now add to these general observations a few remarks about the localities of chromite.

The largest octahedral crystals have been found near Media, Delaware county, especially at the Blue Hill and at Hibbard's farm; very brilliant octahedral crystals, rarely showing the edges truncated, are found at Fairlamb's farm; imperfect crystals in actinolite occur at Mineral Hill.

Good octahedral crystals occur in West Goshen township, one mile from West Chester; in Willistown township; and near the Octarara Creek, in Chester county, and at Reynold's old mine, in Lancaster county.

Very rich massive chromite is found in Marple township, Delaware county, and at the Chester and Lancaster county mines, above enumerated.

A very shining, hard, massive chrome ore from the neighborhood of Media on the farm formerly belonging to Walter Green, contains only a very small percentage of chromic oxide, according to my determination only 26.7 per cent., whilst that from Moro Phillips' Mine, in Marple township, contains 51.56 per cent.

The first analysis of American chromites, amongst which is one from Chester county, were made and published by Henry Seybert (Am. Journ. Sc. IV, 321); octahedral crystals of Chester county chromite was subsequently analyzed by Isaac Starr, and the massive ore from the Wood's Mine in Lancaster county, by Thos. H. Garrett (Am. Journ. Sc. [2] XIV, 45).

Very carefully selected perfectly pure octahedral crystals from Hibbard's farm were analyzed by myself.

		Chester Co Ore.	Chester Co. Oı Octahedral crystals.	Wood's Mine Ore. Massive.	Hibbard's Chromite crystals.
		SEYBERT.	STARR.	GARRETT.	GENTH.
Sp. gr.	=	——	——	4.568	4.780
Chromic oxide	=	51.56	60.836	63.384	53.36
Alumina	=	9.72	0.928	——	5.98
Ferric oxide	=	——	38.952	38.663	7.41
Ferrous oxide	=	35.14	——	——	26.64
Niccolous oxide	=	——	0.100	2.282	0.14
Cobaltous oxide	=	——	——	——	trace.
Manganous oxide	=	——	——	——	0.39
Magnesia	=	——	——	——	6.53
Silicic acid	=	2.90	0.619	——	——
		99.32	101.435	104.329	100.45

Evidently there exists no reliable analysis of any of the various varieties of chromite in Pennsylvania. In those which have been made no attention has been paid to the state of oxydation of the iron, and the separation of alumina. I was anxious, therefore, to make at least one *new* analysis for this Report, and selected Hibbard's, the most beautiful variety in the whole State.

10. *Uraninite* (*Dana*).

This mineral, generally called "pitchblende," and consisting in its pure state of uranic oxide in combination with uranous oxide, has been observed by Mr. Theo. D. Rand in small quantities in granitic veins, in gneiss in the neighborhood of Chester, Delaware county.

4

41. *Rutile* (*Werner*).

It is found in tetragonal crystals, usually consisting of several prisms, octahedra, zirconoids and other planes, frequently in twins, forming so-called "geniculated" crystals.

The forms of the Pennsylvania rutiles have not yet been studied.

Consists of pure titanic acid, usually with a small admixture of ferric oxide. H = 6.65; sp. g. = 4.2. Color reddish brown, inclining to red; powder yellowish brown. Mostly in crystals, rarely massive.

Beautiful varieties of rutile occur in Pennsylvania, the largest and finest crystals in quartz and loose in the soil in Sadsbury township, Chester county, for seven miles along the valley, especially near Parkesburg, where sometimes doubly geniculated crystals of one pound in weight have been found; also in Bart township, Lancaster county.

Fine acicular crytals of rutile in a crystal of limpid quartz found near Kintzer's, Lancaster county, are in the possession of Mr. Joseph Wharton.

In Lancaster county it is occasionally met with in limestone at the Pequea Mine.

Very fine, slender, striated, quadratic prisms with octahedral terminations, sometimes of 2 inches in length and $\frac{1}{8}$ of an inch in thickness, occur in granular quartz, in York county, where also geniculated crystals have been met with.

Splendid acicular crystals with terminations have been found in dolomite and quartz at the Poorhouse Quarry, West Bradford township, and at David Nevin's Quarry, New Garden, Chester county; fine crystals, combinations of the first and second prism with octahedral terminations sometimes four inches long, occur in quartz and limestone in Edward's Quarry, near Doe Run, and near Logan's Quarry, West Marlborough township.

Good crystals have also been found at Wm. Jackson's and Pusey's Quarry, London Grove township, on Brinton's farm, in Thornbury township, the "Black Horse," East Bradford township, and sometimes in the chrome sand washings in West Nottingham township, Chester county.

In Delaware county it has been observed in acicular crystals, in Concord township, in good crystals and massive in quartz in Middletown, Edgemont, Birmingham and Darby townships.

A considerable quantity of rutile has been collected at the principal localities, where it is found loose in the soil and worked up by manufacturers of dentist's supplies for coloring artificial teeth.

42. *Hausmannite* (*Haidinger*).

This rare tetragonal mineral, a combination of manganic and manganous oxides $= MnO,Mn_2O_3$ is reported from Lebanon, Pa. (Dana's Mineralogy 162.) I have not seen it.

43. *Braunite* (*Haidinger*).

This is another tetragonal manganese mineral, probably manganic oxide $= Mn_2O_3$, which has been observed by Mr. Theo. D. Rand, in small quantity at Edgehill, Montgomery county.

44. *Brookite* (*Levy*).

This rare mineral, the monoclinic form of titanic acid, has also been found in this State.

The finest, although imperfect crystal of iron-black color and presenting prismatic and several octahedral planes has occurred at Carter's Chrome Mine, Lancaster county. It is about three-quarters of an inch broad, three-quarters of an inch high, and a little over three-eighths of an inch thick. A fragment proved on examination to be pure titanic acid.

A few slender crystals which appear to be rhombic in form and have been found in chrome sands from the washings of West Nottingham, Chester county, are probably brookite; they have not been examined.

Brookite is mentioned as having occurred with dolomite, quartz, etc., at the railroad tunnel near Phœnixville, Chester county.

45. *Pyrolusite* (*Haidinger*).

Orthorhombic, usually in small rhombic, modified prisms of a dark steel-gray color and strong metallic lustre; in crystalline masses of a blackish or dark gray color and granular. In its pure state, binoxide of manganese, containing 63.3 per cent. of manganese and 36.7 per cent. of oxygen.

It occurs in small rhombic crystals in geodes, frequently associated with turgite in limonite beds, especially in Williams and Saucon township, Northampton county, also under similar circumstances in Berks county and both as pyrolusite and so-called velvet manganese, at Edgehill, and near Spring Mill, Montgomery county.

It is already mentioned (besides psilomelane and wad) by Zach. Cist (Am. Journ. of Sc. IV, 39), as occurring imbedded in a black earthy mass of greasy feeling, the surface covered with minute globules of brilliant lustre on Tobyhannah, Broad Mountain.

b. Hydrous oxides.

46. *Hydrocuprite* (*Genth*)—a new mineral.

Peculiar orange-colored coatings associated with cuprite and magnetite from Cornwall, were first noticed by Prof. W. Th. Roepper, of Bethlehem, who showed them to me several months ago. On a visit to the mines, a short time ago, I secured a considerable number of specimens from the "big hill" at Cornwall, and proved them to be a new mineral.

Amorphous; orange yellow to orange red; forms very thin, sometimes raglike coatings upon magnetite; soft.

On heating loses water and becomes black; contains water and cuprous oxide. It is impossible to obtain from the quantity which I have noticed, enough for analysis; its composition is probably H_2O, Cu_2O.

Besides the amorphous coatings, the cuprite variety *chalcotrichite* from Cornwall, sometimes assumes an orange yellow color, so that on the same piece acicular crystals of a fine crimson color can be seen gradually changing to an orange yellow. It is therefore very probable that the orange crystals are pseudomorphs of hydrocuprite after cuprite.

47. *Turgite* (*Hermann*), or *Hydrohematite* (*Breithaupt*).

In fibrous and divergent crystalline coatings, botryoidal and stalactitic; also massive. $H = 5$; sp. gr. $= 4.14$–4.68. Dark reddish black, powder red. Lustre submetallic, the fibres slightly silky. The surface of the botryoidal concretions is frequently iridescent, but usually black and very lustrous. When heated, it yields water and decrepitates remarkably. It contains 94.7 per cent. of ferric oxide and 5.3 per cent. of water $= 2\ Fe_2\ O_3\ ,\ H_2\ O$.

It has been observed as a coating from the thickness of a varnish to over half an inch upon many of the limonites; most beautiful in those of Lehigh and Northampton counties, also in several of the mines in Berks county, for instance, at Topton, the Moselem Mine, &c.; at the Yellow and Chester Springs, in Chester county; Edgehill, Montgomery county and probably many others.

We have no analysis of a Pennsylvania turgite; Prof. Roepper proved his mineralogical determination by showing specimens from the Lehigh Valley to contain 5.37 per cent. of water.

It has not yet been found independently in massive deposits but only in association with other iron ores.

48. *Diaspore* (*Haüy*).

Crystallizes rhombic, usually in prisms modified by several octahedra and domes; shows an eminent brachydiagonal cleavage; also massive foliated. $H = 6$; sp. gr. $= 3.3–3.5$. Very brilliant; colorless and yellowish or brownish white. Lustre vitreous, but strongly pearly on the cleavage planes.

Consists of alumina 85.1 and water $14.9 = Al_2\ O_3, H_2O$. Decrepitates strongly when heated, yielding water.

This rare mineral has been observed at the corundum locality, near Unionville, in Newlin township, Chester county, in most magnificent crystals, perhaps the best known in the world. Some from a half to one and a half inches in length are perfect, being completely surrounded by planes—others are implanted in corundum; it also occurs in large cleavage masses, up to nearly four inches in diameter. It has generally a slightly brownish color. The best specimens which I have seen are in the cabinets of Dr. Isaac Lea and Col. Joseph Willcox, of Philadelphia, and Mr. W. W. Jefferis, of West Chester.

It has been analyzed by S. P. Sharples (Dana's Mineralogy), who found:

Alumina	=	80.95	per cent.
Ferric oxide	=	3.12	"
Silicic acid	=	1.53	"
Water	=	14.84	"
		100.44	

49. *Göthite* (*Lenz*).

Crystallizes in forms similar to diaspore, in prismatic crystals, cleavage perfect brachydiagonally; fibrous, foliated, scaly; massive, reniform, stalactitic. H = 5–5.5; sp. gr. = 4–4.4. Color from yellowish brown to reddish brown and blackish brown; powder yellowish brown. It contains ferric oxide 89.9, and water 10.1 = Fe_2O_3,H_2O.

It has been observed in minute crystalline coatings, upon limonite, at Edge Hill, Montgomery county; also at the Yellow Springs and Phœnixville, Chester county.

The variety "lepidocrocite" (Ullmann) occurs in scaly fibrous reniform masses, with a fine scaly radiated structure, associated with cobaltiferous wad, upon limonite, at Glendon and other localities in Williams township, Northampton county; also near Yellow Springs, Chester county; at Chestnut Hill, Lancaster county, and at some of the iron mines near Spring Mill, Montgomery county.

The so-called velvet-iron ore has been observed by Mr. H. W. Hollenbush, of Reading, six miles west of Reading, at Sinking Springs.

It probably occurs at many other limonite deposits, but in the absence of analyses it is impossible to state whether many of the so-called "brown hematites" belong to this species or to limonite.

50. *Limonite* (*Beudant*).

The crystalline form of this mineral is not known. It usually occurs in fibrous, radiating, stalactitic, botryoidal and mammillary masses; in concretions, compact and earthy. H=5–5.5; sp. gr.=3.4–4. Its colors are various shades of brown, from dark hair-brown to yellowish-brown, down to brownish-yellow and yellow, the surface often of a black lustrous appearance; the lustre of the fibrous varieties is silky and sometimes submetallic; the massive varieties dull and earthy. When pure it consists of 85.6 ferric oxide, and 14.4 water = $2Fe_2O_3,3H_2O$, but generally containing, even in the apparently purest specimens, silicic acid, phosphoric and arsenic acids, alumina, manganic oxide, etc.

This is the most important iron ore in the State, and deposits of it of greater or less extent have been found in the gneissic rocks of the southern part, also as the results of decomposition of chrysolite rocks in the serpentine range.

Deposits which belong to the gneissic rocks and serpentines are found in Montgomery, Delaware, Chester, Lancaster and York counties.

It shows its greatest development in the lower Silurian limestones, from the Delaware River, in Northampton county, across the whole State, through Lehigh, Berks, Lebanon, Dauphin, Cumberland and Franklin counties; also in Kishicoquilis Valley, Mifflin county; and Brush, Penn's, Sugar, Nittany, Sinking and Canoe Valleys, in Centre and Blair counties; Morrison's Cove, in Blair and Bedford counties, and McConnelsburg Cove, in Fulton county.

Baker and Blair's great limonite banks, near Altoona, Blair county, occur between the lower Helderberg group and the Oriskany sandstone.

Important limonite beds, which may have resulted from the alteration of carbonate of iron or siderite, which at some places has been found in its unchanged state, are worked in Juniata Valley, Mifflin county, and the Great Aughwick and Woodcock Valleys, in Huntingdon and Bedford counties.

Limonite beds are also found overlying the ferriferous limestone of the lower coal measures.

Many of the hematites of the so-called fossil ore beds have by the absorption of water changed to a greater or less extent into limonites.

After having given this general outline of the occurence of the great Pennsylvania iron ore, I will mention a few localities where interesting varieties have been met with.

Very fine fibrous limonite associated with the compact varieties occurs in the serpentine range, at Newlin, East and West Nottingham townships, in Chester county; also near Media, Delaware county; in Lancaster county near Wood's and Low's Mines.

Fibrous, botryoidal, and reticulated varieties, also the so-called "pipe ore" are found near Yellow and Chester Springs, Pikeland township, and at General Trimble's Mine, in E. Whiteland township, in Chester county.

Very fine fibrous limonite occurs at Huntingdon, Blair and Centre counties, and in many of the mines of Northampton, Lehigh and Berks counties; also near Spring Mill and Edge Hill, Montgomery county.

A peculiar variety of pipe ore, consisting of hollow cylinders, with a stem of ore through the middle, occurs at Dr. Thomas' quarry, Whiteland township, Chester county.

A very singular dark brown limonite, with subconchoidal fracture and pitchy lustre occurs with fibrous and compact ores at John Smedley's, Middletown township, Delaware county. (See analysis below, 4.)

Numerous analyses have been made of Pennsylvania limonites, but for the reason already stated I refer to these, and add only a few new ones.

The occurrence of cobalt in limonite was first noticed by M. H. Boyé (Proc. Am. Phil. Soc. IV, 238), who discovered it in a variety from the Chester Ridge, three quarters of a mile west of Chester Furnace, Huntingdon county; it has since been observed in many of the limonites, especially in those which are manganiferous.

Besides the numerous analyses of average ores from Messrs. Lyon, Shorb & Co's ore banks in Centre, Blair and Huntingdon counties (Proc. Am. Phil. Soc. XIV, 84–96), I have examined some apparently very pure limonites from the same localities: 1. Fibrous pale hair-brown limonite from the "Bull Bank;" 2. Dark brown fibrous limonite from the "Dry Hollow Bank;" 3. Pipe ore from the Pennsylvania Bank.

The pitchy limonite from Middletown, Delaware county, in the Laboratory of the University of Pennsylvania, by Mr. Sydney Castle (4), and an apparently pure dark, liver-colored compact limonite from the Wheatley Mine, has been analyzed by J. L. Smith (*l. c.*) (5).

The results are as follows:

	1	2	3	4		5
Ferric oxide =	81.48	83.13	83.74	74.57		80.32
Manganic oxide =	0.07	0.15	0.31	2.58	Oxide of copper =	0.94
Cobaltic oxide =	——	——	trace	0.60	" " lead	1.51
Alumina =	0.49	0.74	0.33	1.54		——
Magnesia =	traces	0.09	0.34	——		——
Lime =		trace	trace	——		——
Phosphoric acid =	0.08	0.50	0.14	0.13		——
Silicic acid =	3.98	2.47	2.57	6.90		3.41
Quartz =	——	——	0.44	——		
Water =	13.90	12.92	12.13	13.10		14.02
	100.00	100.00	100.00	99.42		100.21

In the paper above quoted, I have referred to the peculiar fibrous form, in which the silicic acid remains, when the mineral is dissolved in acids, and have suggested that it may be present in combination with ferric oxide and water, probably as anthosiderite—2 Fe_2O_3, 9 SiO_2 + 2 H_2O.

51. *Brucite* (*Beudant*).

Crystallizes in the hexagonal system, usually in combinations of several rhombohedra with the basal plane; cleavage eminently basal, foliæ easily separable; also foliated massive; rarely fibrous. H = 2.5; sp. gr. = 2.35. Lustre between vitreous and waxy, on the cleavage face pearly; colorless, white, yellowish, grayish, blueish, greenish and pink.

When pure it contains: Magnesia 68.97 per cent., Water 31.03 per cent. = MgO,H_2O.

This rare mineral occurs in Pennsylvania in magnificent specimens, far more beautiful than in any other part of the world; the greatest variety of forms has been obtained from Wood's Mine, Lancaster county; not only numerous modifications and very large tabular crystals, sometimes several inches in diameter and of very delicate shades of color have been observed, but also groups of crystals in rosettes and globular masses, and the foliated mineral in abundance; it occurs rarely in radiating masses inclining to fibrous.

It is also found at Low's Mine, but the specimens are not equal to those from Wood's Mine.

In small quantity it has been observed near Unionville, Newlin township, Chester county.

We have analyses of it from the two principal localities, that from Wood's Mine, by Hermann (Journ. für Prakt. Chemie, LXXXII, 368); that from Low's Mine, by Smith and Brush (Am. Journ. of Sc. [2] XV, 214), who found:

	Wood's Mine.	Low's Mine.
Magnesia	68.87	66.30
Ferrous oxide	——	0.50
Manganous oxide	0.80	trace
Water	30.33	[31.93]
Carbonic acid	——	1.27
	100.00	100.00

52. *Gibbsite* (*Torrey*).

Usually in mammillary, stalactitic incrustations, having a faint fibrous structure; also in minute hexagonal crystals, with strong pearly lustre upon the basal plane, or in delicate concretionary crusts. Various shades of white. The pure mineral contains: Alumina 65.6 per cent., water 34.4 per cent. = $Al_2O_3,3H_2O$.

It has been observed by T. F. Seal (Am. Journ. of Sc. [2] XI, 267) as a thin coating and small mammillary incrustation upon albite, near Unionville, Newlin township.

Also upon limonite, associated with wavellite on General Trimble's farm, near White Horse Station, on the Chester Valley Railroad. The latter variety is grayish, pearly subtranslucent, coating the limonite in thin delicate concretionary crusts of H = 3, and sp. gr. = 2.35; it has been analyzed by Hermann (Bull. Soc. Impér. Nat. Moscow No. 4, 1868, 493), who found:

Alumina	=	63.84
Water	=	33.45
Silicic acid	=	1.50
Phosphoric acid	=	0.91
Magnesia and Ferric oxide	=	traces
		99.70

It is also reported from the iron mines at Chester Springs, West Pikeland township, Chester county.

53. *Psilomelane* (*Haidinger*).

Not crystallized, usually in stalactitic, botryoidal, mammillary masses, of an iron-black color, inclining to blueish black or steel-gray. Lustre submetallic, also dull; opaque. H = 5–6; sp. gr. =3 .7,–4.7.

The minerals, which are called psilomelane are in their purest state mostly combinations of manganese binoxide, with manganous oxide, baryta, potash or even lithia.

Usually it contains mechanical admixtures of other minerals, such as pyrolusite, manganite, limonite, &c., and it is often impossible to distinguish this species and the massive varieties of pyrolusite, except by analysis.

It is a frequent associate of limonite, and has been noticed in many of the mines of Northampton, Lehigh, Berks, Blair, Centre, Huntingdon, &c., counties.

It has been found in gneiss, at the Pennypack Creek, three miles from Bustleton, Bucks county.

Zach. Cist described in 1821 a variety of a manganese ore, which evidently belongs to psilomelane: compact indurated manganese, blue and purple-black, very heavy, fracture splintery, in detached masses from the size of a walnut to a man's head—on the headwaters of Bear Creek, Lehigh and Tobyhannah, Broad Mountain.

54. *Wad* (*Kirwan*).

This includes the soft and earthy varieties of hydrated oxides of manganese which are frequently called "*Bog Manganese*;" other varieties containing, besides the oxides of manganese, considerable quantities of cobalt, have been named *asbolite*; and the cupreous varieties have been distinguished as *lampadite*; the color of the latter two is usually black or brownish-black.

It has been observed as a frequent associate of limonite ores. Most of the wad or bog manganese of the limonite region of Northampton, Lehigh, Berks, Montgomery and other counties, contains some cobalt.

A large deposit of an ochreous cobaltiferous variety occurs, according to Prof. Roepper (priv. com.), near Albertis, Lehigh county, at the mine of the Philadelphia and Reading Coal and Iron Company.

A cobaltiferous wad, containing, according to Mr. Theo. D. Rand, 10 per cent. of cobalt, has been observed by him in the lowest stratum of the drift opposite Fairmount, West Philadelphia.

Some varieties of the Lehigh region contain a considerable admixture of clay, as is shown by the analysis of one by Prof. W. Th. Roepper (priv. com.).

He found:

Clay	=	45.66	per cent.
Ferric oxide	=	19.29	" "
Manganic oxide	=	23.64	" "
Magnesia	=	0.34	" "
Phosphoric acid	=	0.29	" "
Water	=	11.04	" "
		100.26	" "

A black soft mineral, which is dull, but when cut becomes shining and of a somewhat waxy lustre and is rarely found imbedded in chrysocolla, at Cornwall, has been partially examined by myself, and contained manganic oxide, cobaltous oxide and cupric oxide in the ratio: 70: 40: 44.

It may be a cobaltiferous *crednerite.*

---

2. *Oxides of Elements of the Arsenic and Sulphur Groups.—Series II.*

55. *Molybdite.*

This mineral is found in yellowish white and yellow earthy incrustations upon quartz, and associated with molybdenite, from the oxydation of which it is formed, at Upland on the Chester Creek in Delaware county.

3. *Oxides of the Carbon-Silicon-Group.—Series II.*

56. *Quartz.*

There is no mineral in Pennsylvania equal in importance to quartz, an essential element in the construction of many of the rock-masses throughout the State.

Together with muscovite it forms the mica slates or mica schists; and with muscovite and feldspar, usually orthoclase, the gneiss and granite, which are so well developed in the south-eastern portion of Pennsylvania.

After the breaking up of the rock-masses of the older formation and the disintegration and frequently the complete decomposition of the micaceous and feldspathic constituents into clays and other minerals, the less destructible quartz, ground down to sand and pebbles, gives rise to sedimentary rocks, consisting of sandstones, conglomerates, &c., from the Potsdam sandstone up to the most recent formations.

As it lies not in the province of this report to treat of these, I shall confine myself to a brief outline of the more interesting occurrences of the numerous varieties throughout the State.

Quartz crystallizes in the hexagonal system, the usual form being a combination of two rhombohedra, often making hexagonal pyramids, mostly with the hexagonal prism. There are a great many other forms of less frequent occurrence. Also massive, granular, fibrous, radiating. H = 7; sp. gr. = 2.5–2.8. Colorless, white, and various shades of yellow, gray, red, green, blue, brown, black. Lustre vitreous, in some varieties resinous; splendent to dull. Fracture, conchoidal, sub-conchoidal and uneven. When pure: Silicic acid = SiO_2.

This mineral occurs in many varieties:

a. Crystalline varieties.

a. The colorless variety or *rock crystal* has been found in many localities:

Very interesting specimens have been found in the dolomite of the Poorhouse Quarry, West Bradford, Chester county. They are mostly hexagonal prisms with the pyramid, frequently distorted; some larger crystals upon a smaller hexagonal stem;

some perfectly colorless, others, especially those of larger size, say four inches in length and from one to one and a-half in thickness, not clear.

Good quartz crystals have been found near Pennsbury, Chester county, especially at Swain's; the almost microscopic quartz crystals, which are inclosed between plates of muscovite from this locality form very beautiful groups.

Fine transparent crystals, some one and a half inches in length and half an inch thick, occur at Crystal Spring, on Blue Mountain, in Bushkill township, Northampton county.

Quartz crystals with good terminations are found on Thos. Lee's farm in Oley township, and small colorless crystals in the Crystal Cave, near Kutztown, Berks county.

Transparent crystals, sometimes doubly terminated and highly modified, but mostly short and stout, from one and a half to three inches in size, and opaque on the base but colorless on the hexagonal pyramid, and showing a distinct and easy rhombohedral cleavage, are found abundantly in the soil near Nazareth, Northampton county.

It is also found in good colorless crystals at Broad Mountain; at Shamokin; at Cherry Valley, and Poconac Valley, Monroe county; at Pequea, New Holland and Kintzer's, in Lancaster county; at Frankstown, Huntingdon county.

Limpid quartz in crystals is found in the soil five miles from Philadelphia; also in East Whiteland, Chester county, and many other localities.

Large rough crystals doubly terminated, some weighing over 15 lbs., have been met with at London Grove; similar ones, occur at King of Prussia, Montgomery county, also clusters and groups of doubly terminated fine crystals between this place and the Schuylkill; crusts made up of quartz crystals are found in Maginnis' lime quarries, Montgomery county. Good crystals have been obtained from the tunnel and the mines near Phœnixville, and good druses of quartz crystals at Cornwall, Lebanon county.

Yellow crystallized quartz occurs at Birmingham, and brown quartz in small crystals and porous at Middletown, Delaware county.

Especially in the serpentine region, a variety of quartz, which has been called "drusy quartz," is very common, having been eliminated by the decomposition of the original chrysolite and its alteration into serpentine. It is usually found as an incrustation made up of very minute crystals, often ferruginous and frequently graduating into ferruginous quartz and also into chalcedony. It has been found, amongst other localities, in Concord, Marple, &c. townships, in Delaware county; at Brinton's Quarry and in Newlin, East Bradford and West Goshen townships, Chester county, and at Wood's Mine, Lancaster county.

Radiated quartz occurs at Diamond Rock, near Warren's Tavern, Chester county; stellated quartz at Perkiomen, Montgomery county, and an interesting variety of fibrous quartz (which had been considered petrified horsehair), with zincblende and limonite, at the zinc mines at Friedensville, near Bethlehem.

b. *Amethyst*, color blueish violet or purple, always in crystals, usually the hexagonal prism terminated with the pyramid

Magnificent specimens of this variety have been found at many localities of Delaware and Chester counties; however, the color is usually not uniform and deeper at the apex, and therefore stones, which could be used as gems, are rare.

Fine specimens have occurred at S. Entrikin's, Wm. Gibbon's, Mrs. Foulke's and Dr. Elwyn's farms, in East Bradford township, also at John Entrikin's and Jos. B. Darlington's farm, Pocopson township, and in Birmingham township, Chester county. Smaller crystals, of a lighter purple color, also in clusters have been obtained from Sadsbury Village, Chester county.

Splendid crystals, one weighing seven pounds, have been found at Morgan Hunter's farm in Upper Providence, also in Lower Providence, Aston, Concord, Marple, Middletown townships; near Twaddle's paper-mill, in Birmingham; in clusters and isolated crystals near Dutton's mill, Chester and Thornbury, Delaware county.

Well crystallized amethyst has also been found at New Salem, about seven miles from Gettysburg, Adams county.

A very peculiar variety of amethystine quartz has lately been observed at Shaw and Ezra's Quarry, near Chester, Delaware county. The crystals frequently inclose smoky quartz upon which they have crystallized, and in juxtaposition with each other are arranged so as to form large plates of crystals sometimes a foot in diameter, six to eight inches in height and half to two inches in thickness. The color of this amethyst is smoky purplish.

c. Smoky quartz. From smoky gray and brown and transparent to brownish-black and opaque.

Smoky quartz crystals have been found near Philadelphia on the Schuylkill; near Reading, Berks county; near Hummerstown, Dauphin county; in Delaware county, in Upper Darby, near Garrett's Road Toll-gate, and near the Kellyville Schoolhouse; at the tunnel near Phœnixville; and in East Nottingham and Birmingham townships, Chester county.

Some of the quartz crystals from Poorhouse Quarry are smoky on the top; others from Upland, Delaware county, are between amethyst and smoky quartz.

Near New Holland, Lancaster county, a variety of smoky quartz, not of a very good color, however, has been found in crystals, from four to five inches in length and about two inches thick.

Massive smoky quartz occurs in Sadsbury township, Chester county.

d. Milky quartz has been obtained from Sadsbury township, Chester county.

e. Blue quartz occurs very fine in its massive variety at Geo. Van Arsdale's Quarry, Bucks county; it has also been found at Radnor, Delaware county and in East Bradford and Uwchlan township, Chester county.

f. Green quartz. It is found in small but beautiful isolated crystals, twins and crystalline groups of a pale green color at Blue Hill, and in inferior specimens near Dismal Run, Middletown, Delaware county.

g. Cat's-eye has been observed in several localities; a fine hexagonal crystal with the pyramid, of greenish color, resulting from very fine fibres of actinolite disseminated through it came from York county; it is also found five miles east of Bethlehem at the allanite locality.

h. Aventurine quartz has been found at Conshohocken.

i. Ferruginous quartz in aggregations of small brownish

yellow crystals and in granular masses is found in the limonite mines of Saucon Valley; it is also found at Mineral Hill, Delaware county.

k. Fetid quartz is found at Mendenhall's limestone quarry in Pennsbury, Chester county; and the zinc mines at Friedensville, Lehigh county.

A peculiar granular laminated quartz rock resembling *itacolumite* has been found at Parksburg, Chester county.

Common quartz occurs in some localities of Delaware and Chester counties in large quantities, and is mined for the manufacture of silicate of soda, sandpaper, &c.

β. Crypto-crystalline varieties.

a. Chalcedony.

Usually in botryoidal, mammillary and stalactitic masses or coatings on other minerals; various shades of color from white, blue, grayish, yellowish, brownish, to dark-brown or black.

The drusy quartz of the serpentine range frequently graduates into chalcedony; we find it therefore at many of the localities already mentioned.

It occurs in Middletown, Marple, and other townships in Delaware county; at the Hopewell Mine, Warwick township, Chester county; in brown botryoidal masses in Willistown; also in West Nottingham and West Goshen, and in botryoidal forms in London Grove townships, Chester county; in Lancaster county it occurs near Rock Springs and Wood's Mine; in Berks county, between Oley and Hamburg, also at Flinthill; and in Monroe county in Cherry Valley; in Montgomery county, it has been observed at Conshohocken; a pale blue variety has been found near Gettysburg, Adams county and a pisolitic variety at Cornwall, Lebanon county.

b. Carnelian or the *red* and brownish-red variety of chalcedony has been observed at Marple, Delaware county, and the Rock Springs, Lancaster county.

c. Prase, a translucent leek-green variety of chalcedony and quartz occurs in the syenitic range of the Lehigh, especially at the allanite locality, five miles east of Bethlehem.

d. Agate, a variegated chalcedony, has been found in Marple township, Delaware county, and the so-called "moss-agate" at

Rock Springs, Lancaster county, and near Reading, Berks county.

e. Agate-jasper occurs at several places in the neighborhood of Reading, Berks county, graduating into several other varieties such as, jasper, hornstone, &c.

e. Flint, somewhat allied to chalcedony, but more opaque and of gray, brownish and grayish-black colors, has been found in rolled pieces in the gravel hills on the Schuylkill, and of black color in the Delaware.

f. Hornstone. Resembles flint, but is usually more brittle and in its fracture more splintery.

A flint-like variety of hornstone is very abundant in the lower strata of the Lower Silurian limestone, graduating in many localities into a jet black, very fine-grained, compact basanite.

Dr. Leidy remarks that a variety from the limestone of Easton reveals an oolitic composition when examined under the microscope with low power. (Proc. Ac. N. Sc. Phil. 1867, 125.)

Basanite is found in the drift of the Delaware River from Easton down to the State line.

A white hornstone is found at Wood's Mine, near Texas, Lancaster county; a brownish variety with blackish striations near West Chester, in West Goshen township, Chester county; other localities are the neighborhood of Philadelphia, on the Wissahickon, Conshohocken and many others.

g. Jasper. An impure opaque colored quartz with dull or waxy lustre.

There are many places in Pennsylvania where specimens which might be called jasper have been found; it is abundant in Berks county, near Reading and in the drift of the Schuylkill and Delaware Rivers.

Brownish-yellow jasper occurs in West Goshen township, Chester county, and a reddish-brown variety near Texas, Lancaster county; a brown-banded jasper is found at the hydropathic establishment near Bethlehem.

The arrow-heads which are found in the neighborhood of Bethlehem and Easton are mostly made of jasper.

h. I have yet to mention some interesting specimens of

"*pseudomorphous quartz*," which have been found in Pennsylvania.

It occurs in flat rhombohedral crystals after calcite near the Gap Mine, in Bart township, Lancaster county.

At Cornwall, Lebanon county, it is found mostly in the same form, but also in hexagonal prisms, which, however, may be combinations of a very acute rhombohedron and basal plane.

Pseudomorphous quartz has been found in an old quarry five miles from Philadelphia, between Old York and Germantown Roads.

Neither variety of crypto-crystalline quartz occurs in the State in sufficient quantity and beauty to be of technical value.

57. *Opal* (*Plinius*).

Massive, amorphous, some varieties in small reniform and botryoidal coatings or stalactites, also earthy. H = 5.5-6.5; sp. gr. = 1.9–2.3. Colorless, white, yellow, red, brown, green, blue, gray. Composition, when pure: silicic acid = SiO_2, *in the amorphous state*. Usually containing a considerable quantity of water and foreign matters.

Opal is a rare mineral in Pennsylvania, and only two varieties have been observed.

The common opal occurs rarely at Cornwall, Lebanon county, in small masses of a grayish, greenish and yellowish white color, and vitreous lustre, inclining to resinous.

It has been examined in the Laboratory of the University of Pennsylvania, by Dr. Wm. P. Headden, who observed that almost the whole was soluble in dilute potash solution. Sp. gr. = 1.879. It contains:

Silicic acid	=	92.65
Ferric oxide and Alumina	=	1.57
Water	=	5.55
		99.77

The variety *hyalite* is occasionally met with in yellowish fluorescent coatings (probably colored by uranic oxide, as suggested by Mr. Theo. D. Rand) upon gneiss at Frankford; also at Megargee's Paper-mill, on the Wissahickon, and in blueish green coatings (probably colored by cupric oxide) upon gneiss at Avondale, Delaware county.

A peculiar substance from the serpentine barrens near West Chester, which has been called silicate of magnesia, probably belongs to opal. It is fine-grained massive, of a white and grayish-white color, rough to the touch, and consists almost wholly of silicic acid. It has not yet been fully investigated.

B. Ternary Oxygen Compounds.

I. *Silicates.*

a. Anhydrous Silicates.

1. *Bisilicates.*

Oxygen ratio for bases and silicic acid = 1 : 2.

58. *Enstatite* (*Kenngott*).

Orthorhombic; crystals rare; usually in crystalline masses, the individuals showing a distinct cleavage of a prism of 87° and 93°. Occurs in two varieties, the magnesian, which is characteristic for some meteoric stones; and the *ferriferous* or *bronzite*, which latter occurs at several localities in Pennsylvania, and has a grayish to brownish-green and brown color and shows on the cleavage planes an adamantine, pearly lustre, inclining to sub-metallic. H = 5.5 ; sp. gr. = 3.3.

Its composition is $(MgFe)O,SiO_2$.

It is often erroneously called "anthophyllite."

It has been found in Rose's serpentine quarry opposite Lafayette, in Montgomery county; both in lustrous cleavage masses and more or less altered into serpentine.

It occurs at many localities in Delaware county. I have observed it near Crump's Serpentine Quarry, near Media in Middletown township; and largely developed in Marple township near Henry Hipple's; and forming the mass of the "Castle Rock" in Newtown township.

Dr. George Smith (*l. c.*) mentions additional localities of the laminated or stratified variety in Radnor, and near the old limekiln in Newtown townships.

A beautiful massive foliated variety occurs under similar circumstances, half a mile west of Texas, Lancaster county.

The Pennsylvania bronzite has several times been analyzed. An analysis of a variety from Leiperville (means probably Delaware county, as it does not occur in the *immediate* neighborhood of Leiperville) by Pisani is given in Descloizeau's Mineralogie I, 537; the Texas mineral has been analyzed by Th. H. Garrett (Am. Journ. Sc. [2] XV, 333).

They found:

		Pisani.	Garrett.
Silicic acid	=	57.08	55.45
Alumina	=	0.28	1.13
Ferrous oxide	=	5.77	9.60
Manganous oxide	=	——	0.98
Magnesia	=	35.59	31.83
Water	=	0.90	——
		99.62	98.99

Bronzite appears to be an important rock in its connection with the serpentine range.

The association of bronzite with chrysolite is very common. Not only most of the granular chrysolite boulders (olivine), which are imbedded in basaltic rocks, invariably contain bronzite, but also many rocks in Norway, Moravia, Lake Lherz in the southern part of France, and nearer home, in North Carolina, Georgia and Alabama, show the same association.

It was only a few weeks ago that I made the very interesting observation that the bronzite rock from Castle Rock, Delaware county, contains grains of chrysolite disseminated through the whole mass.

There was hardly time yet to make an investigation of this subject, but I thought it of importance to verify my observations by a few preliminary analyses at least, which were made at my request in the Laboratory of the University of Pennsylvania, by Mr. Pedro G. Salom.

He found in the portion soluble in dilute chlorhydric acid, being between 5 and 10 per centum of the whole rock:

Silicic acid	=	45.15
Ferrous oxide	=	19.38
Magnesia	=	31.49
Lime	=	3.88
Alumina	=	trace
		99.90

which leaves, no doubt, that the greater part of the soluble portion is *chrysolite.*

The part insoluble in acids contained:

Silicic acid	=	55.89
Ferrous oxide (and Alumina)	=	12.05
Niccolous oxide	=	0.65
Magnesia	=	29.15
Lime	=	1.22
		98.96

which closely agrees with the composition of *bronzite.*

59. *Wollastonite* (*Haüy*).

Monoclinic; rarely in crystals; usually in cleavable masses; fibrous. H = 4.5–5; sp. gr. = 2.8–2.9. White, grayish and yellowish-white. Lustre vitreous to pearly.

Silicate of lime, containing when pure: Silicic acid 51.7; lime 48.3.

It has been found in crystalline tabular masses of fibrous structure and of fine pearly lustre at Geo. Van Arsdale's Quarry in Bucks county, and contains according to the analysis of S. G. Morton (Journ. Ac. Nat. Sc. Phila. VI, 46).

Silex (silicic acid)	=	51.50
Lime	=	44.10
Ferrous oxide	=	.00
Ignition	=	0.75
		97.35

The so-called *wollastonite* from Easton is tremolite.

60. *Pyroxene* (*Haüy*).

This mineral is found in many varieties, few of which have been observed in Pennsylvania.

Crystallizes in the monoclinic system with a prism of 87° 5′, in combination with about fifty different planes. Cleavage is perfect parallel to the prismatic planes and, nearly so, parallel to the basal and orthodiagonal. The crystals are usually thick and stout; frequently in lamellar masses, also granular and fibrous. H = 5–6; sp. gr. = 2.23–3.5. Color, various shades of an impure green, graduating to greenish-gray and white, and on the other side through greenish-black to black. The general formula of pyroxene is RO,SiO_2; RO being variable quantities of lime, magnesia, ferrous and manganous oxides (in a few varieties zinc oxide and alcalies); some contain also alumina and ferric oxide.

As none of the specimens from Pennsylvania (with one exception) have been analyzed, they are placed under the name of such varieties which they resemble most closely.

Alalite (?) or *Mussite* is found in grayish-white columnar, cleavable masses in Bailey's limestone quarry, East Marlborough township, Chester county.

Mussite in minute white crystals associated with magnetite from Cornwall, Lebanon county, is in Prof. Roepper's collection.

It occurs also at Brinton's Ford on the Brandywine in Pennsbury township, Chester county.

Sahlite occurs in small crystals of various shades of green, but more frequently in cleavable masses, or granular, so-called *coccolite*, disseminated through calcite and wollastonite at Geo. Van Arsdale's Quarry, near Feisterville, Bucks county; *coccolite* has also been observed at Steele's Mine, near St. Mary's, and Brinton's Ford, Chester county; it accompanies allanite, near Bethlehem, and occurs frequently in calcite and the white limestones in the neighborhood of Easton and Bethlehem.

The variety sahlite is found in large masses in the South Mountain range and a greenish massive variety southwest of Shimersville had, some forty years ago, been mistaken and worked for an iron ore.

It appears to be an essential constituent of the so-called "syenite" of the South Mountains together with a feldspar, which gives by analysis the composition of orthoclase, some portions of which, however, show triclinic striations.

Fragments of this sahlite, which have been selected from a

piece of this rock from the Philadelphia and Reading Coal and Iron Company's Mine at Seissholtzville, near Albertis, in Lehigh county, were analyzed in the Laboratory of the University of Pennsylvania, by Mr. Sydney Castle, who found:

Silicic acid	=	49.30
Alumina	=	14.98
Ferric oxide	=	0.53
Ferrous	=	6.02
Manganous	=	trace
Magnesia	=	8.27
Lime	=	21.45
		100.00

Diallage, in dark greenish laminated masses, has been found near Unionville, Newlin township, Chester county.

Augite is mentioned (Dana's Mineralogy) as occurring near Knauertown, Warwick township; at Chrisman's lime quarry, South Coventry township, and near Brinton's Ford, Pennsbury township, Chester county.

The pyroxene crystals from the Gap Mine, Lancaster county, mentioned in Dana's Mineralogy, are hornblende.

A grayish-green crystal, which seems to be a pyroxene, has been found in a micaceous rock near Texas, Lancaster county.

A great deal is to be done yet to learn the exact places where some of the pyroxene varieties belong.

Probably many of the so-called "trap rocks" of Pennsylvania contain augite as an essential constituent. It will require, however, a careful investigation of these rocks by the examination of microscopic sections and chemical analysis, to ascertain their true constitution.

61. *Amphibole* (*Haüy*).

Many varieties of amphibole or hornblende have been found in the State, but the composition of two only has been determined by analysis.

Amphibole crystallizes in monoclinic prisms, with an angle of 124° 30′; numerous combinations have been observed. Cleavage highly perfect parallel to the prismatic planes. Usually in crystalline masses, fibrous and columnar, the individuals often very long and fine, like flax; also granular, massive; usually tough, some varieties brittle. H = 5–6 ; sp. gr. 2.9–3.4. Colors between white and black, many varieties grayish, brownish, greenish-white and green. Lustre vitreous and pearly on cleavage planes, often silky in the fibrous varieties.

The chemical composition is the same as that of pyroxene.

Tremolite occurs in beautiful crystals, of white color, with perfect terminations in John Bailey's limestone quarries and near Doe Run Village, in West Marlborough township; of snow-white color in fine acicular crystals and fibrous, and in radiated masses, in the limestone quarries near Avondale, New Garden, London Grove and Kennett townships, Chester county; it has also been found near the old limekiln, on the West Chester road in Newtown, Delaware county, and at Chestnut Hill, near Easton.

A variety, which occurs in thin, long, bladed crystals in oligoclase on Painter's farm near Dismal Run, Middletown township, Delaware county, probably belongs to this variety.

Actinolite is that variety of amphibole, which is most abundantly found in Pennsylvania.

I include under this name the varieties which are generally called anthophyllite or antholite, numerous graduations from the typical actinolite into tremolite on one side and into cummingtonite on the other.

It is frequently met with in the serpentine range of Delaware and Chester counties.

The typical actinolite of a pale green, yellowish and brownish-green or grass-green colors has been found abundantly at Mineral Hill, Middletown township; also in Concord, Upper Providence and Marple townships, where the radiated variety occurs in talc; and in greenish crystalline masses at Rockdale. It is also found in Willistown, Kennett and Newlin townships, Chester county, and a *yellow* variety in talc at Caleb Cope's lime quarries in East Bradford.

Actinolite in chlorite slate is found near Waggontown, Chester county, and in green compressed crystals in talc on the serpentine ridge, Little Britain township, Chester county.

In Concord township, Delaware county, occurs the so-called antholite, a grayish and brownish variety, radiated and stellated, some of the rays six and a half inches in diameter; the analysis of this is given below; it is also found in Aston, Marple, Middletown and Upper Providence townships in the same county.

The very beautiful variety of actinolite "byssolite" is found in calcite and with magnetite at Knauertown, and very fine globular crystalline nodules, radiating from a centre, at the St. Mary's Iron Mines, Warwick township, Chester county; also at Cornwall, Lebanon county.

Actinolite occurs also at Easton and in acicular crystals and silky fibres half a mile from Bustleton, Bucks county.

In irregularly grouped acicular crystals of fine green color in talcose rocks it is found on the Wissahickon, and at the Columbia Bridge near Philadelphia.

The emerald green, glassy actinolite of Concord township, Delaware county, has been analyzed by H. Seybert (Am. Journ. Sc. VI, 331) and the so-called "antholite" from the Star Rock, Concord township, Delaware county, by A R. Leeds (Am. Journ. Sc. [3] VI, 26):

		Seybert.			Leeds.
Spec. gr.	=	2.987			3.20
Silicic acid	=	56.233			55,12
Alumina	=	1.666			0.55
Ferrous oxide	=	4.300			8.20
Magnesia	=	24.000			31.18
Lime	=	10.666			0.75
Chromic oxide	=	trace	Manganous oxide	=	0.33
Water	=	1.033	Soda	=	1.55
		———	Potash	=	1.01
		97.998	Water	=	2.21
					100.90

Asbestus, in its hard ligniform variety, with silky lustre, the fibres sometimes broken across and separated by seams of quartz, also a soft fibrous and pliable variety of a yellowish white color, occur near Rockdale, in Aston township, Delaware county, where it has been mined to some extent for the manufacture of pulp and felt for covering boilers, &c. It is found in other localities in the same county, at Mineral Hill, in Middletown, in Upper Providence, Radnor, &c., townships.

A very excellent quality of asbestus is found near Moro Phillips's Chrome Mine, West Nottingham township; it also occurs in East Nottingham; at Strode's Mill, East Bradford; in East and West Goshen, Chester county; the soapstone quarries and Rose's serpentine quarries near Lafayette, Montgomery county, and at Chestnut Hill, near Easton, and an inferior variety at the Gap Mine, Lancaster county.

Very delicate fibres of asbestus with crystallized quartz are found in a quarry of hornblendic gneiss on the Schuylkill, four miles from Philadelphia.

The so-called "mountain leather" of a brownish color is found upon the limestone in the zinc mines at Friedensville, near Bethlehem; it is also found in the limestone quarries near Avondale and the Poorhouse, Chester county; in Newlin township, Chester county, a somewhat fibrous variety of "Mountain cork" has been met with.

Hornblende in greenish-black crystals and radiating masses is found in the iron mines at Knauertown, Warwick township, Chester county; a dark greenish fibrous and radiating variety occurs at the soapstone quarries, near Lafayette, Montgomery county.

Crystals of a dark green color and about one inch in length by half an inch in thickness, are found at the Gap Mine, Lancaster county.

A massive crystalline granular hornblende of a greenish-black color, near Rockdale, Delaware county, has been mistaken and mined for "coal."

Black hornblende occurs abundantly as a constituent of hornblendic gneiss in the southeastern portion of Pennsylvania.

In many localities of this whole range it becomes frequently void of other constituents, and the hornblendic gneiss graduates into hornblendic schists and hornblende rock.

S. P. Sharples (Am. Journ. Sc. [2] XLII, 271) has made an analysis of a crystalline very dark green hornblende rock, forming a bed of from one to two miles long and half to three-quarter miles wide in Birmingham township, Delaware county. He found:

Silicic acid	=	47.77
Alumina	=	7.69
Ferrous oxide	=	15.41
Manganous oxide	=	0.26
Magnesia	=	15.28
Lime	=	13.16
		99.57

Hornblende is also a constituent of many trap rocks, usually invisible to the naked eye; their structure, as already stated, can only be established by microscopic and chemical investigation.

62. *Crocidolite* (?) (*Hausmann*).

A fibrous, asbestus-like mineral of silky lustre and a lavender and blackish-blue color, principally a silicate of ferrous oxide, magnesia and soda.

Mr. Theo. D. Rand observed a dark-blueish fibrous mineral at the Falls of Schuylkill, which appears to be this mineral; and Prof. W. Th. Roepper found associated with white and brownish-white garnet, at Coopersburg, blueish-white crystalline fibrous coatings which may belong here.

Neither has been examined.

63. *Beryl.*

Hexagonal prisms with basal plane and several hexagonal and dihexagonal pyramids. Cleavage indistinct. H = 7.5–8; sp. gr. = 2.63–2.75. Color from emerald-green into pale green, yellowish and grayish-green and white. Transparent to translucent; brittle.

Beryl is a silicate of alumina and glucina, $= 3\,BeO,SiO_2 + Al_2O_3,3\,SiO_2$, containing silicic acid 66.8, alumina 19.1 and glucina 14.1.

Throughout the southeastern gneiss region of Pennsylvania, especially in the granitic segregations, beryl is frequently met with; Delaware county especially is rich in interesting localities.

Fine crystals of almost emerald-green beryl have been found associated with small crystals of black tourmaline in a dark micaceous schist at Deshong's quarry, near Leiperville; in the granitic veins of the same quarry and in the vicinity very beautiful slender crystals of yellowish, greenish and blueish beryl have been found; they are sometimes ten to twelve inches in length and one and a half inches in diameter, and frequently terminated with the basal plane.

Very beautiful crystals, often doubly terminated with pyramids and basal plane and of a yellowish-green color have been met with at Shaw and Ezra's quarry, near Chester.

A crystal from the same locality, in Dr. Cardesa's cabinet, seems to be an acute hexagonal pyramid.

Large hexagonal prisms in the same collection are from Wm. Trainer's farm, at Upland; yellowish-green crystals sometimes four inches in length by one and a half in diameter, but much broken by fissures, are found in Springfield township; Upper Providence, Middletown, Concord and Marple townships have also furnished fine specimens; at the White Horse, three to four miles below Darby, it is found in twelve-sided prisms in blue cyanite.

In Chester county it occurs in six and twelve-sided prisms, sometimes of a large size (one crystal in the cabinet of Mr. Wm. W. Jefferis, weighing 51 lbs.), and of a blueish-green and blue color in the neighborhood of Unionville, in Newlin township; and on the Brandywine battle-ground in Birmingham township.

In greenish-white hexagonal prisms in granular quartz it occurs at Amos Pugh's farm, East Nottingham township; and in a granite deposit in Brinton's serpentine quarry, in doubly terminated hexagonal prisms in quartz.

A white variety in well defined crystals from one to two inches in length and a quarter to one inch in diameter has rarely been found on the Old York Road, five miles from Philadelphia; at the same place occasionally occur yellow and pale green beryls, sometimes of pure rich and uniform color; opposite Fairmount Water Works it is found in yellowish-green crystals, sometimes altered; also elsewhere in West Philadelphia; and at Flat Rock Tunnel on the Philadelphia and Reading Railroad.

None of the varieties are of sufficient purity of color and transparency to be fit for gems.

None of the Pennsylvania beryls have been analyzed.

2. *Unisilicates.*

Oxygen ratio for bases and silicic acid = 1 : 1.

64. *Chrysolite* (*Wallerius*).

This important mineral has almost completely disappeared from Pennsylvania, the great mass formerly existing having been altered into talc and serpentine, which occupy now such a prominent position in the southeastern part of the State (Amer. Jour. of Soc. [2] XXXIII, 199).

Orthorhombic; prism 94° 2′. Usually in crystalline, granular masses or imbedded grains. H = 6–7; sp. gr. = 3.5. Color, various shades of green, usually yellowish, grayish or brownish-green; vitreous.

Its chemical composition is a silicate of magnesia and ferrous oxide, $2(MgO.FeO),SiO_2$.

A fragment of a crystal, part of which is in the cabinet of the University of Pennsylvania, and the other part in that of Dr. Cardesa, of Claymont, Delaware, was found many years ago at Wood's Mine, Lancaster county. It has a yellowish-green color and a vitreous lustre and is associated with chromite, kaemmererite, &c.

A few months ago I found grains of chrysolite imbedded in the bronzite-rock of Castle Rock in Newtown township, and also in Middletown township, Delaware county.

It is probable that it occurs in some of the crypto-crystalline trap rocks in the State.

65. *Garnet* (*Albertus Magnus*).

Isometric in crystallization. The usual forms are the dodecahedron and trapezohedron. There are several other forms, neither of which, I believe, has been observed in Pennsylvania. Usually in crystals, also in crystalline and granular masses. H. = 6.5–7.5; sp. gr. = 3.15–4.3. Colors, white, yellowish, brownish, red, brown, black, green. Lustre vitreous, some varieties inclining to resinous.

The general formula for garnet is:
$3\ (2\ RO,SiO_2) + 2\ R_2O_3,3SiO_2$, in which RO may be lime, magnesia, manganous oxide and ferrous oxide; R_2O_3, alumina, ferric oxide and chromic oxide.

But one of the Pennsylvania garnets has been analysed, therefore some doubt exists, to which some of the varieties belong.

a. *Lime-alumina-garnet.*

To this variety belong, probably the following: the white or brownish white garnet, occurring in small dodecahedral crystals and crystalline aggregations near a trap dyke at

Coopersburg; the green garnets found near Hummelstown, Dauphin county, and the grayish green granular garnet. from Fritz's Island Mine near Reading.

b. *Iron-alumina-garnet.*

Most of the garnet, which has been found in the State, belongs to this species.

The granitic segregations of the gneissic rocks of the south-eastern part, especially in Delaware county, have furnished many magnificent specimens, mostly of a brownish red and reddish brown, sometimes almost black color, and in dodecahedral and trapezohedral crystals, frequently combinations of both, one or the other predominating.

Near Bishop's Mills, in Middletown township, very dark reddish brown crystals of truncated dodecahedra, from three to four inches in diameter, are found in orthoclase; very fine reddish brown trapezohedral garnets, the larger ones with dodecahedral planes, some two and one-half inches in diameter are found at Avondale, in Springfield township; also in Concord township; at Deshong's Quarry, Shaw and Ezra's Quarry, and Upland near Chester; it is also found in Darby, Aston, Lower Providence, Haverford and Radnor townships.

A rock, consisting largely of red granular garnet occurs on the James Lancaster farm.

In Chester county, beautiful trapezohedral garnets of a brownish-red color are found near Unionville, in Newlin township; trapezohedral garnets from one-half to one inch in diameter, sometimes in muscovite, and compressed between the mica-plates occur near Fairville, Pennsbury township; dodecahedral garnets are found in the soil near M. Phillips's Chrome Mine, West Nottingham township; trapezohedral and dodecahedral garnets occur also in East and West Goshen, London Grove, Penn, Lower Oxford, Elk and East Bradford townships.

In the neighborhood of Philadelphia, the iron-alumina garnet is found in dodecahedra, with trapezohedral planes, back of Germantown; and in dodecahedral crystals abundantly on the Wissahickon; in the soapstone quarries on the Schuylkill near Lafayette; in rounded dodecahedral crystals in the soil on Frank Fennimore's farm on Gulf Creek; at Flat Rock Tunnel and the Falls of the Schuylkill. One very beautiful crystal, five inches in diameter, has been found at Barren Hill meeting house, twelve miles from this city.

Granular garnet and small dodecahedral crystals of a pale red color are disseminated through blue quartz and calcite at Van Arsdale's, Bucks county.

Brown garnet associated with sphene is found five miles East of Bethlehem in Lower Saucon township, Northampton county.

A dark-red variety similar in color to pyrope, is found in the bed of Darby Creek near the Lazaretto in Delaware county.

Very peculiar garnets are found above Peters' mill-dam in Green's Creek, Concord township, Delaware county. Some have a deep blood-red color and have been mistaken for pyrope; that they are *true* garnets has been proved by an analysis made many years ago in my Laboratory, by Mr. Chas. A. Kurlbaum, Jr. (Am. Journ. Sc. [2] XIX, 20).

Dr. J. Leidy (Proc. Ac. Nat. Sc. Phil., 1871, 155) states that he never found them in crystalline form, and that they show a singularly corroded appearance, probably from the abstraction of other minerals imbedded therein; the garnets probably existed in the rock in the form of nodules and not in crystals.

Dr. Isaac Lea (Proc. Ac. Nat. Sc. Phil., 1869, 4) in his "Notes on microscopic crystals in minerals," observes that of 310 specimens examined, 75 contained acicular crystals inclosed.

Kurlbaum's analysis gave:

Silicic acid	=	40.15
Alumina	=	20.77
Ferrous oxide	=	26.66
Manganous oxide	=	1.85
Magnesia	=	8.08
Lime	=	1.83
		99.34

Some of these specimens have a fine color, and when cut make a beautiful gem.

c. *Manganese-alumina-garnet.*

It is found abundantly in brownish-red, finely granular masses, one-quarter of a mile west of Ridge Road and nine miles from Philadelphia, also abundantly in the soil between Germantown turnpike and Roxboro' township line, six and a-half miles from Philadelphia.

Massive manganese garnet also occurs in Concord and Middletown townships, Delaware county, and on Osborne's Hill, East Bradford township, Chester county; the massive garnet found on V. Hartman's farm in Alsace township, near Reading, probably belongs to this variety.

d. *Lime-iron-garnet.*

Very brilliant brownish-black, and sometimes jet-black, large dodecahedral crystals with truncated edges, associated with a feldspathic mineral, or smaller crystals, from a quarter to over one inch in diameter, occur in geodes at St. Mary's; similar crystals, some over two and a-half inches in diameter, have been found at Keim's Mine, near Knauertown; also dark-brown garnets at Steele's Mine, near St. Mary's, all in Warwick township, Chester county.

Melanite is reported as occurring at Warren.

Very brilliant black dodecahedral garnet, from half an inch to one inch in diameter, occur on White's farm, south of Gettysburg, in Adams county, near the Maryland line.

They all belong, probably, to this variety.

The Knauertown garnets are often erroneously called "aplome," but as they do not show the least trace of brachydiagonal striation, they do not belong to this variety.

e. *Ouvarowite or Lime-Chrome-garnet.*

This most beautiful and rarest of all garnets was first noticed in 1866, by Prof. C. U. Shepard (Am. Jour. Sc. [2] XLI, 216), as occurring in minute, nearly transparent, emerald-green crystals not over one-tenth of an inch in size, scattered through a pale green, granular clinochlore, intermingled with brownish-gray vermiculite, at Wood's Mine, Lancaster county; it has since been found but once by Mr. Th. D. Rand, who has a small emerald-green crystal of it, showing some of the rhombic planes of the dodecahedron very distinctly.

66. *Zircon* (*Klaproth*).

Crystallizes in the tetragonal system. The quadratic prism of first and second order with several quadratic and octagonal pyramids (so-called "zirconoids"), are the usual forms. Cleavage, imperfect. Usually in crystals. H = 7.5 ; sp. gr. = 4–4.75. From colorless to yellowish, grayish and brownish-white, yellowish and reddish-brown, brown and black. It consists of silicate of zirconia = silicic acid 33 ; zirconia 67.

Very fine clove-brown quadratic prisms with pyramids from one-sixth to two inches in length and one-tenth to one-half inches in breadth have formerly been found imbedded in the talc of Chestnut Hill, near Easton.

Associated with allanite in small crystals of the usual form, rarely half an inch in length and partly altered on all the planes, except the zirconoid, the color of the latter being brownish-black, while the others are gray, it is found in Lower Saucon township, Northampton county, five miles east of Bethlehem ; and under similar circumstances, three-fourths of a mile north of Bethlehem.

It is found in minute slender crystals in the South Mountains one mile east of Hellertown.

Rarely it occurs in small prismatic crystals, with two pyramids and the zirconoid, at Van Arsdale's, Bucks county.

It is found imbedded in magnetite near Pricetown, about ten miles east-northeast of Reading, in chocolate-brown, opaque crystals, seldom as large as one and a half inches by three-eighths of an inch. The crystals are rarely perfectly terminated, the pyramidal and zirconoid planes have their edges rounded, as if subjected to fusion. They have been analyzed by Dr. Chas. M. Wetherill (Trans. Am. Phil. Soc. X, 340), who found:

Spec. gr.	=	4.595
Silicic acid	=	34.07
Zirconia	=	63.50
Ferric oxide	=	2.02
Water	=	0.50
		100.09

Good crystals are also found on the farms of W. Hains and rarely on Schroeder's and Mrs. D. Rhodes, half a mile east of Pricetown, Berks county.

In Chester county it is found in crystals in iron ore half a mile from South Coventry Village, and at Chester Springs and Yellow Springs, West Pikeland township ; brown crystals of zircons are found loose in the soil at Pusey's sawmill, two miles southwest of Unionville in East Marlborough township; also in quartz, in East Bradford about two and a half miles west of West Chester; near the paper-mill on the Brandywine, very fine prismatic crystals of an inch in length are found associated with (?) menaccanite in blueish quartz.

In small crystals it occurs in quartz and a triclinic feldspar in the stones on the turnpike, Willow Grove, to the top of Camp Hill, fourteen miles from Philadelphia.

It was formerly found (Dr. Troost, in Journ. Ac. Nat. Sc. Phil. II, 57, March 6th, 1821) in crystals of a quarter of an inch in length—combinations of the first and second prism with two pyramids and the zirconoid—on the York Road, fifteen miles from Philadelphia. Crystals with predominant zirconoid planes have been found by Benj. Say, at the same place.

A most beautiful variety, in minute brown crystals—prisms terminated by the zirconoid—occur with oligoclase on Painter's farm, near Dismal Run, Middletown township; it has also been observed at the Blue Hill, in Upper Providence township, Delaware county.

The chrome sand of Chester and Delaware counties contain sometimes small, but beautiful crystals of zircon, with the zirconoid predominating.

The gravel of the Delaware and Schuylkill Rivers contains a considerable quantity of very minute, nearly colorless crystals of zircon.

67. *Vesuvianite* (*Werner*), *Idocrase* (*Haüy*).

Tetragonal; usually in tetragonal and octagonal prisms with numerous pyramids and zirconoids and the basal plane. Cleavage indistinct. H = 6.5; sp. gr. = 3.35–3.45. Colors between brown and green, sometimes yellow. Lustre vitreous. Composition not fully established, perhaps $6(2RO,SiO_2)+2R_2O_3,3SiO_2$, RO being principally lime, with smaller quantities of magnesia, ferrous and manganous oxides; R_2O_3 = alumina, sometimes partly replaced by ferric oxide.

This is a rare mineral in Pennsylvania.

A perfect beautiful wine-colored variety has been found near New Hope, Bucks county (Roger's Geol. Rep. Pa. II, 685).

Imperfect crystalline masses, showing a striation, similar to that frequently observed with some varieties of vesuvianite, are occasionally found at Cornwall, Lebanon county; they are probably vesuvianite.

Columnar greenish, yellowish, brownish imperfect crystalline aggregations are rarely found at Wood's Mine. They have a close resemblance to vesuvianite.

None of these varieties has been analyzed

68. *Epidote* (*Haüy*).

Epidote crystallizes in the monoclinic system. Crystals often very complicated, and usually lengthened in the direction of the orthodiagonal; cleavage perfect orthodiagonally; also columnar, fibrous, divergent, granular, compact. H. = 6–7; sp. gr. = 3.25–3.5. Colors green, from pistacchio to yellowish and brownish-green; lustre, vitreous; fracture uneven, brittle. The composition of epidote is $3(2RO,SiO_2) + 2(2R_2O_3,3SiO_2)$, RO being principally lime and ferrous oxide; R_2O_3, alumina with ferric oxide.

Epidote has been found at many localities in Pennsylvania, often in connection with hornblende or pyroxene rocks, and probably, in part at least, as the result of their alterations.

Chester county furnishes fine varieties in beautiful bottle green prismatic crystals, from one-half to three inches in length, and one-tenth to three-fourths of an inch in thickness; it has been found on Smith and McMullin's farm in West Bradford township; in East Bradford township it is found in crystals of a dark green color, also in brownish crystals in quartz, and at Strode's Mills in yellowish-green prismatic crystals; it has also been observed at R. Taylor's Mill, West Goshen; in East Marlboro township it occurs in yellowish green crystals, in cavities of hornblende rock, and in a similar association at Wm. Cloud's farm and Pearce's old mill, in Kennett township; in bottle-green crystals it has been observed in limestone quarries of London Grove township; it also has been met with in Sadsbury township.

Near Philadelphia it is found in crystals and crystalline masses of a dark bottle-green color in a calcite vein in gneiss at Frankford; on the Falls of the Schuylkill and on the Wissahickon.

Small crystals occur on feldspar one mile east of Reading, in Berks county, and abundantly on the South Mountains in greenish-yellow crystals and granular masses, as one of the products of the alteration of pyroxene rocks, for instance, with allanite, three-fourths of a mile north, and five miles east of Bethlehem; near Easton it is found in small dark green crystals and compact; in a similar manner in Bucks county, near New Hope.

Massive and fibrous epidote occurs with native copper at several localities near Gettysburg, in Adams county.

None of the Pennsylvania varieties has been more fully investigated either crystallographically or chemically.

69. *Allanite* (*Thomson*).

Monoclinic and isomorphous with epidote. Perfect crystals are rare, and none have been found in Pennsylvania. Massive. H = 5.5–6; sp. gr. = 3–4. Colors, dark pitch-brown to black: powder greenish brown. Lustre resinous, inclining to vitreous.

Essentially a silicate of alumina and ferric oxide, with ferrous oxide, cerous oxide, lanthana, didymia and lime.

It has been found in a decomposed gneiss in East Bradford township, on Amor Davis' farm, about one mile from West Chester, mostly in the soil from three to four feet below the surface, in pieces from the size of a grain to six inches in length and of one pound weight. The pieces are mostly covered with a thick brown earthy crust of the decomposed mineral. The analysis by Prof. Rammelsberg (Poggendorff's Annalen, LXXX, 285) is given below.

It has also been observed near Pughtown, Coventry township, Chester county.

Associated with magnetite and zircon it occurs east-northeast of Reading, near Pricetown.* It is of a pitch-black color and massive; some of the specimens, however, show rough crystalline planes, and are also somewhat coated with decomposed earthy mineral. It has been analyzed in my laboratory by Peter D. Keyser (Am. Jour. Sc., [2] XIX, 20); another interesting variety has been found in the South Mountains,

* This is usually published as "from Eckhardt's Furnace," from which place the first specimens were obtained.

near Bethlehem, associated with zircon, sphene, &c., the principal locality being five miles east of, and another three-fourths of a mile north of Bethlehem. It occurs in a granitic rock in narrow veins or seams in large imperfect crystals and crystalline masses of a blackish-brown color, decomposed on the surface with a yellowish-brown earthy crust. The analysis, also by P. D. Keyser (*l. c.*), is given below.

A large mass of allanite of about one hundred pounds in weight has lately been found on Lehigh Mountain, just south of Lehigh University.

The analyses gave the following results:

		East Bradford.	Priceton.	Bethlehem.
Sp. gr.	=	3.535	3.831	3.491
Silicic acid	=	31.86	32.89	33.31
Alumina		16.87	12.49	14.34
Ferric oxide		3.58	7.33	10.83
Ferrous oxide		12.26	9.02	7.20
Manganous oxide		——	0.25	——
Cerous oxide		21.27	15.68	13.42
Lanthana, Didymia }		2.40	10.10	2.70
Lime		10.15	7.12	11.28
Magnesia		1.67	1.77	1.23
Soda		——	0.09	0.41
Potash		——	0.14	1.33
Water		1.11	2.49	3.01
		101.17	99.37	99.06

70. *Zoisite* (*Werner*).

Orthorhombic; combinations of numerous prisms, octahedral planes and domes. Perfect crystals rare, usually much lengthened in the direction of the vertical axis, frequently deeply striated. Cleavage very perfect, brachydiagonally; also in crystalline masses, radiating, columnar. H = 6–6.5; sp. gr. = 3.11–3.38. The usual colors are white, grayish-white, brownish-white, and brown. Lustre vitreous to pearly. The composition is: $3(2CaO,SiO_2) + 2(2Al_2O_3,3SiO_2)$.

The purest variety has been found as the product of the alteration of corundum, at Unionville, Newlin township, Chester county, it had been called "Unionite," (B. Silliman, Jr., Am.

J. Sc., [2] VIII, 384). It is implanted in black tourmaline, in crystalline masses, showing one distinct cleavage dividing the mineral into parallel laminæ.

It has been analyzed by Geo. J. Brush (Am. Jour. Sc., [2] XXVI, 60), whose analysis is given below.

It was found when constructing the West Chester Water-works, in a hornblendic gneiss, in which it occurs in radiating, somewhat columnar masses, with distinct cleavage, and of a grayish-brown color. This locality is now inaccessible.

In a similar manner it occurs in Kennett township, Chester county.

It has formerly been found in hornblendic gneiss on the Schuylkill, near Philadelphia, in acicular crystals and fascicular grayish masses of a pearly lustre.

It is found associated with hornblende and albite at Prince's Soapstone quarries near Lafayette, Montgomery county, both in minute indistinct grayish and brownish crystals and crystalline masses. I have made an analysis of the latter.

The analyses of the Pennsylvania zoisites gave the following results:

		So-called "Unionite." BRUSH.	Soapstone Quarries. GENTH.
Sp. gr.	=	3.299	——
Silicic acid	=	40.61	40.30
Alumina	=	33.44	28.31
Ferric oxide	=	0.49	5.47
Manganous oxide		——	0.47
Magnesia		trace	0.66
Lime		24.13	22.93
Water		2.22	1.86
		100.89	100.10

The so-called *zoisite* from Darby township, Delaware county, is fibrolite.

71. *Ilvaite* (*Steffens*).

Mr. Th. D. Rand observed this species in very small black crystals with carbonate of lime and in a narrow vein in gneiss, at Flat Rock Tunnel on the Phila. and Reading Railroad, opposite Manayunk.

It is an orthorhombic mineral, essentially a silicate of ferric oxide with ferrous oxide and magnesia.

Mica Group.

The members of this group, of which only three have been observed in Pennsylvania, are alike in having a prismatic angle of 120° ; they have an eminently basal cleavage, affording very thin, tough laminæ ; their crystallization is either hexagonal or rhombic and therefore the optic-axial plane at right angles to the cleavage ; they all contain potash and alumina as essential constituents.

From the external characters without optical and chemical examination it is frequently impossible to determine to which species a specimen may belong, and as but very few of the Pennsylvania varieties have been investigated, it is often doubtful where to place them.

72. *Phlogopite* (*Breithaupt*).

Orthorhombic, habit hexagonal, usually in oblong six-sided prisms, with irregular sides; cleavage basal, highly eminent. H. = 2.5—3 ; sp. gr. = 2.85. Color yellowish-brown to brownish-red, green, white and colorless. Lustre pearly and submetallic. Transparent; optic-axial divergence 3°–20°, rarely less than 5°. Mostly a silicate of alumina, magnesia and potash ; some varieties contain considerable quantities of ferrous oxide, soda and fluorine.

It occurs rarely in Pennsylvania ; in brown plates with graphite, pyrrhotite, &c., at Van Arsdale's quarry, in Bucks county.

In small brownish, crystalline plates in Bailey & Brothers' limestone quarries, East Marlborough township, and Nivin's lime-quarries, New Gardon township, Chester county.

No Pennsylvania phlogopite has been examined optically except a brownish-olive green variety from Warwick, resembling much the Vesuvian biotites, with an optical angle of over 10°.

A dark brown mica, which occurs in small plates in the Gap Mine ore may also belong to phlogopite.

There are no analyses of phlogopites from this State.

73. *Biotite* (*Hausmann*).

Hexagonal, habit often monoclinic; prisms usually tabular; cleavage basal, highly eminent. Often in disseminated scales and massive aggregations. H. = 2.5–3; sp. gr. = 2.7–3 1. Colors usually dark green, brown to black, rarely white. Uni-axial, but from exceptional irregularities sometimes bi-axial with slight axial divergence, the angle not exceeding 5°, seldom 1°. The chemical constitution is not well understood. Biotites are mostly silicates of alumina, part of the latter sometimes replaced by ferric oxide, with magnesia, ferrous oxide and potash.

In grayish and greenish-white hexagonal crystals, in part altered into a serpentine-like mineral; also in large foliated masses of silver-white color and sub-metallic lustre at Easton. These appear to be undoubted biotites, they have been optically examined by Grailich, who finds the axial divergence for the silvery-white 1°–2°; and for the green 3°–4°.

A black, dark brownish or greenish-black mica, very common in the gneissic and granitic rocks of Southeastern Pennsylvania, and frequently inter-crystallized with muscovite, either the black mica in the muscovite or the grayish-white muscovite in the black mica, is biotite.

It is sometimes partly altered into a hydrous mineral, which exfoliates on heating, probably jefferisite.

Senarmont has found that a clear olive-green mica from Philadelphia gives an axial divergence of 3°–4°.

The principal localities where it is found are in West Philadelphia, from the junction railroad north of Girard avenue bridge to Gray's Ferry; at Painter's Cross-roads and Stamp's tavern in Concord and Middletown townships and the vicinity of Chester, Delaware county; also at Abner Marshall's in Pennsbury township, Chester county.

A biotite from Penneville, Pa. (perhaps Pennsville, Chester county), has been analysed by Ed. F. Neminarz (Tschermak's Miner. Mittheilungen 1874, 241). It had a brown color and sp. gr. = 2.779.

The analysis gave:

		Pennsville. NEMINARZ.
Fluorine	=	1.94
Silicic acid	=	44.29
Alumina	=	12.12
Ferric oxide	=	1.40
Ferrous oxide	=	1.44
Magnesia	=	27.86
Lithia	=	trace
Soda	=	2.16
Potash	=	7.06
Water	=	2.09
		100.36

It is probably a frequent constituent of many mica schists and gneissic rocks, which often abound in black or very dark mica in minute scales; but as they have not been investigated, this cannot yet be stated with certainty.

74. *Muscovite* (*Dana*).

Orthorhombic with monoclinic habit. The prismatic angle is 120°. Usually in hexagonal crystals, with eminent basal cleavage. Folia often aggregated in stellate, globular and plumose forms, also scaly massive. H. = 2–2.5; sp. gr. = 2.75 –3.1. Colors usually white, gray, brown, hair-brown, brownish-green and green. Thin laminæ flexible and elastic. Optic-axial angle 44°–78°. Muscovite is principally a silicate of alumina and potash, part of the alumina sometimes being replaced by ferric oxide and the potash by soda. Most of the muscovite analyses give water, but it has not yet been fully established whether it belongs to the constitution of the muscovite, or whether the hydrous mineral does not really belong to damourite.

Beautiful and large crystals of a brownish white color, sometimes in plates from twelve to fifteen inches diameter and weighing one hundred pounds, have been found near Fairville, Pennsbury township, especially on the farm of Jacob Swain and Wm. Dilworth; it is also found in long striated plates, sometimes only in thin fibres, at Marshall's farm in the neighborhood.

The muscovite from Pennsbury shows frequently red, blue and green colors, or inclosures of minerals flattened by the mica, such as garnet, and minute crystals of quartz, and, as above stated, magnetite in beautiful groups; muscovite with acicular crystals of tourmaline comes from Thornbury township.

In hexagonal prisms it is found near Unionville, Newlin township, and in globular aggregations, made up of hexagonal plates and radiating, it is found in quartz at Amos Pugh's Mill, East Nottingham township, Chester county.

Good hexagonal semi-transparent crystals, but smaller than those from Pennsbury, are found at Dutton's Mills, and yellowish, greenish hexagonal crystals at Shaw and Ezra's and other quarries near Chester, Delaware county.

In hexagonal crystals, often with bands of reticulated magnetite, it is found in Middletown township; in crystals, inclosing crystals of biotite, and in peculiar imperfect crystals, one-half being perfect and with smooth sharp planes, while the other half is wanting, in Concord township; and several other localities in Delaware county.

Muscovite inclosed in Jefferisite occurs three miles south of Westchester, in Westtown township, Chester county.

Near Philadelphia muscovite is found in good crystals on the Falls of the Schuylkill and in West Philadelphia, from a quarry opposite Fairmount to Gray's Ferry; also near Germantown.

A green mica, which may be muscovite, has been found at Chestnut Hill, near Philadelphia, and in foliated masses and crystallized near Unionville, Newlin township; and one of a leek-green color on Wm. Jackson's farm, in London Grove township, Chester county; also at Van Arsdale's quarry, Bucks county.

No muscovite of Pennsylvania has been analyzed; of a few, the optic-axial angle has been determined. Prof. B. Silliman found it to be as follows:

Pennsbury, smoky brown striated,	59°
Philadelphia, greenish gray banded,	60° 30′–61°
" near Fairmount, smoky brown,	60°–62° 30′
Unionville, white, corundum locality (probably damourite, Genth),	67°–67° 28′

Pennsbury brown crystals—another locality,	69° 27′–70°
" brownish green—third locality,	70°–70° 30′
Leiperville, Delaware county, faint greenish, plicated,	70° 30′–71°
A transparent, clear olive green muscovite, from Philadelphia, measured by Senarmont, gave,	57°–58°

Muscovite is a very important element in the constitution of rock-masses: with quartz alone it forms mica slate or mica schist; with quartz and feldspar (usually orthoclase) granite or, when the rock is laminated, gneiss.

---

75. *Wernerite* (*d'Andrada*).

Tetragonal. The usual forms are combinations of the first and second prism, with a pyramid of 133° 7′. In short, stout crystals, also massive. H = 5–6; sp. gr. = 2.6–2.8. Colors, white, grayish-white, gray. Lustre vitreous, inclining to greasy.

A silicate of alumina, lime and soda, with the oxygen ratio of $RO : Al_2O_2 : SiO_2 = 1 : 2 : 4$.

Crystals of a grayish-white color, showing the prismatic and pyramidal planes, but mostly grouped so as to show no determinate forms, have been found in Geo. Van Arsdale's Limestone quarry, near Feisterville, Bucks county; the massive gray variety occurs very abundantly at the same locality; the translucent columnar, massive variety has been analyzed by Prof. A. R. Leeds (Am. Journ. Sc. [3] VI, 26) who found:

Spec. gr.	=	2.708
Silicic acid	=	47.47
Alumina	=	27.51
Ferric oxide	=	trace
Magnesia	=	1.20
Lime	=	17.59
Soda	=	3.05
Potash	=	1.40
Water	=	1.48
		99.70

It is also reported as occurring near Marshalton, West Bradford; at Nivin's quarry, New Garden; near Doe Run Village, in Kennett, and at Bailey's quarries, in West Marlborough townships, all in Chester county.

Feldspar Group.

There is such a great similarity between several minerals in their physical characters and chemical composition, and such a close resemblance in their mode of occurrence that they are usually classed together under the general name of the group of "Feldspars."

Their general characteristics are given in Dana's Mineralogy nearly as follows:

They have a spec. gravity of less than 2.85; a hardness between 6 and 7; a monoclinic or triclinic crystallization; the prismatic angle is near 120°; two easy cleavages, one basal, the other brachydiagonal, inclined at an angle of 90° or very near 90°. Cleavage a prominent feature of many massive kinds and distinct in the granular varieties; close isomorphism and a general resemblance in the systems of occurring crystalline forms; twinning parallel to the clinodiagonal section and O and sometimes 2-ì (or the corresponding triclinic planes); transition from granular varieties to compact, hornstone-like kinds, called felsites; often opalescent or having a play of colors as seen in a direction a little oblique to i-ì, often aventurine from disseminated microscopic crystals of foreign substances parallel mostly to planes O and *I*.

The bases of the species, which occur in the State are lime, soda and potash combined with alumina and silicic acid. The species, which so far have been recognized, including labradorite, which probably exists in some of the trap rocks, are:

	Crystallization:	Oxygen ratio for: RO	: R_2O_3	: SiO_2
Labradorite	triclinic	1	3	6
Oligoclase	triclinic	1	3	9
Albite	triclinic	1	3	12
Orthoclase	monoclinic	1	3	12

These three latter species occur in Pennsylvania in numerous varieties, but of many a great deal of doubt still exists as to whether they belong to one or the other, or may perhaps form distinct species; of a few their true place has been established by chemical analyses.

The crystals which have been found in Pennsylvania have not yet been studied, but the crystalline structure, especially the cleavage of some of the varieties from the neighborhood of Philadelphia, has been investigated by Mr. Theo. D. Rand (Notes on feldspars—Proc. Ac. Nat. Sc. Phil 1872, 299).

In a paper " Notes on some members of the feldspar family," read May 4th, 1866, by Isaac Lea (Proc. Ac. Nat. Sc. Phil. 1866, 110–113) Dr. Lea gives the results of microscopic examinations of sections of many varieties from Chester and Delaware counties and several other localities, and distinguishes several varieties under the names: Lennilite, Delawarite and Cassinite. I hope to be able to furnish the results of analyses of these hereafter.

The feldspars are very predominant constituents of rock masses in Pennsylvania, especially in the southeastern part, where the gneiss and granite are chiefly made up of quartz, mica and orthoclase; albite, and perhaps oligoclase, frequently are found in subordinate quantities.

Albite or oligoclase are often associated in the so-called hornblendic gneiss rocks, and are also not uncommon in association with serpentine.

In the South Mountain Range feldspathic rocks occur, containing besides feldspar, a variety of pyroxene. It is not certain yet what this feldspar is From a striation which I observed on a fragment, it would appear that it was triclinic, while the analysis which Dr. Headden has made gives the composition of orthoclase.

All the so-called "trap rocks" contain a considerable percentage of one or more feldspars.

A full investigation of these crypto-crystalline rocks by a microscopic examination of sections and a chemical analysis is very desirable, as it may throw much light upon many points of interest and importance.

I shall now proceed to give the principal localities of the species, and shall make free use of the mark of interrogation, where it is has not yet been ascertained whether the mineral found at a certain place, belongs to one or the other species.

76. *Labradorite.*

Labradorite has probably not been found in the State in any other but the crypto-crystalline form in the rocks forming trap dykes.

The opalescent feldspars which have been observed at Van Arsdale's, in Kennett township, the Magnesia quarries and near Media belong to other species.

77. *Oligoclase* (*Breithaupt*).

Crystallizes in triclinic forms. Usually in crystalline cleavable masses, often showing striations. Cleavable in two directions, intersecting at angle of 93° 50′. H. = 6–7; sp. gr. = 2.6–2.7. Colors whitish, with a tinge of gray, yellow, green or red. The variety "sunstone" or "aventurine-oligoclase" shows yellow and red reflections from minute inclosed crystals, usually of hematite or goethite. Lustre vitreous, inclining to pearly, also waxy.

A silicate of alumina, soda and lime, giving the oxygen ratio of $RO : Al_2O_3 : SiO_2 = 1 : 3 : 9$. Some varieties inclining towards 1 : 3 : 12 are probably mixtures of oligoclase and albite.

No crystals have been observed in Pennsylvania.

Very beautiful varieties of *sunstone*—which probably belong to oligoclase—occur at Wm. Cloud's farm, three miles south of Kennett Square, and at Pearce's paper-mill, half a mile south of Kennett Square, and in the hornblende rocks at various other places in Kennett township. They consist of small patches sometimes of one to two inches in diameter, which, when examined under a microscope with high power, present a great number of red hexagonal plates, producing beautiful reflections. Associated with the sunstone of Peace's paper mill, a beautiful feldspar has been found by Mr. Wm. W. Jefferis, showing as good an opalescence as the best labradorite; whether it is an oligoclase or a labradorite is doubtful.

Very good sunstone (? oligoclase), with very beautiful reflections, has been found near Fairville, Pennsbury township; sunstone (? oligoclase) occurs also at Mendenhall's Lime Quarries, Pennsbury, Chester county; sunstone (? oligoclase) has also been found in Aston township and on Mineral Hill, near Media, Delaware county; some of it is of a grayish white color with coppery reflections; other varieties are sunstone inclosed in so-called moonstone (albite).

Masses of cleavable oligoclase, with deep triclinic striations, is found with jefferisite in Westtown township, three miles south of Westchester; also at Strode's Mill, East Bradford township, and with tourmaline, euphyllite and corundum at Unionville, Newlin township, Chester county.

A transparent yellowish cleavable oligoclase with deep striæ has been obtained from Painter's farm, near Dismal Run, Middletown township, Delaware county; (see analysis below).

A granular triclinic feldspar from the same township with implanted crystals of corundum, partly altered into fibrolite, is probably also oligoclase.

The lamellar blue feldspar, striated on its surface, from Rob't Lamborn's farm, West Bradford township, Chester county, is probably oligoclase.

The following varieties have been more fully investigated:

1. A white variety from Unionville; it is associated with euphyllite, showing vitreous lustre and a distinct cleavage in one direction, analysis by Smith and Brush (Am. Jour. Sc, [2]XV, 211, and XVI, 44); 2. a granular yellowish or brownish-white variety, the small cleavage planes of which are finely striated, occurring associated with corundum at Unionville, Newlin township, Chester county, has been examined by Mr. Thos. M. Chatard (my paper on Corundum, *l. c.*); 3. at my request Mr. Reuben Haines analyzed in the Laboratory of the University of Pennsylvania the transparent cleavable feldspar with deep triclinic striations from Painter's farm, near Dismal Run, Middletown, Delaware county.

The analyses gave the following results:

		1. White (Unionville). Smith & Brush. (Mean of two analyses.)	2. Brownish-white. (Unionville). Chatard.	3. Dismal Run. Haines.
Sp. gr.,	=	2.61	——	2.68
Silicic acid,	=	64.27	59.35	63.54
Alumina,	=	21.21	24.16	22.96
Ferric oxide,	=	trace.	0.61	0.16
Magnesia,	=	0.58	0.34	——
Lime,	=	0.81	3.08	4.21
Soda,	=	10.94	7.22	8.44
Potash,	=	1.36	3.78	0.59
Ignition,	=	1.08	1.96	0.52
		100.25	100.50	100.42

78. *Albite* (*Gahn and Berzelius*).

Triclinic.—Resembles oligoclase so much that by external appearance it cannot be distinguished from it. In crystals and more frequently in cleavable masses, showing sometimes thin laminations; also granular, occasionally impalpable.

Silicate of alumina and soda, with the ratio of $RO : Al_2O_3 : SiO_2 = 1 : 3 : 12$.

Crystals of albite have rarely been met with in Pennsylvania. Very minute white crystals have been found in Cumberland county;

With zoisite it occurs at the soapstone quarry near Lafayette, in Montgomery county.

It (or ? oligoclase) is found in the gneiss of West Philadelphia, from opposite Fairmount to Gray's Ferry. Rarely it is met with in fine crystals, but abundantly in cleavable masses in the gneissic rocks in the vicinity of Chester, Delaware county. Peculiar aggregations of imperfect crystals, which my analysis proved to be albite have occurred at Shaw & Ezra's quarry, near Chester.

Beautiful specimens of so-called "moonstone" with blue opalescence have been found at Mineral Hill, near Media, in Middletown township, Delaware county. (See analysis.)

The laminated albite, so-called "cleavelandite," has been found with cyanite at Cope's Mills, East Bradford township, and a white granular variety as the matrix of corundum, at Unionville, Newlin township (see analyses).

There is a variety of semi-transparent triclinic feldspar, which is sometimes associated with orthoclase, and partly altered into deweylite, from Unionville, Newlin township; it may be albite or oligoclase; greenish-gray granular albite or oliglocase, rarely showing a blueish opalescence, occurs in serpentine at the magnesia quarries, West Nottingham township, Chester county, and also near Glen Mills, Delaware county, and a grayish-white granular variety of great hardness and toughness in the chrome mines of Lancaster county.

Several varieties of albite from Pennsylvania have been analysed:

1. Densely aggregated crystalline grains, like moderately coarse-grained marble, associated with corundum and other minerals, from Unionville, by Martin H. Boyé (and Booth)

(Proc. Am. Phil. Soc., II, 190); 2. the analysis of the same variety has been repeated by M. C. Weld (Am. Jour. Sc., [2] VIII, 390, in a paper of Prof. Silliman, Jr.); 3. the analysis of the massive compact granular variety of a grayish-white color, resembling dolomite. from a chrome mine in Lancaster county, has been made by Geo. J. Brush (quoted in the same paper of Prof. Silliman); 4. the variety "moonstone," with brilliant blue reflections upon the principal cleavage, and a fine striation upon the second, from Mineral Hill, Middletown township, Delaware county, has been analyzed by A. R. Leeds (Am. Jour. Sc., [3] VI, 25); 5. another variety "from Pennsylvania" has been analyzed by Redtenbacher, (Pogg. Annalen LII, 49); 6. the crystals from Shaw and Ezra's Quarry, Delaware county, have been analyzed by myself.

The following are the results:

		Unionville.		Lancaster county.	Moonstone, Mineral Hill,	Pennsylvania.	Shaw & Ezra's Delaware county.
		1. BOYÉ.	2. WELD.	3. BRUSH.	4. LEEDS.	5. REDTENBACHER.	6. GENTH.
Spec. Grav.	=	2.612	——	2.619	2.59		
Silicic acid	=	67.72	66.857	66.653	67.70	67.20	68.03
Alumina	=	20.54	21.889	20.786	19.98	19.C4	20.23
Ferric oxide	=	trace	——	——	trace	——	trace
Magnesia	=	0.34	0.481	0.519	0.11	0.31	——
Lime	=	0.78	1.785	2.050	1.47	1.44	0.48
Soda	=	10.65	8.779	9.360	8.86	9.91	10.81
Potash	=	0.16	——	——	1.36	1.58	0.09
Ignition	=	——	0.481	——	0.08	——	0.62
		100.19	100.272	99.368	99.56	100.07	100.26

79. *Orthoclase* (*Breithaupt*).

In monoclinic crystals, combination of numerous planes; cleavage perfect on O, i-ì less distinct. Frequently in twins. Often massive, lamellar and granular, also compact crypto-crystalline. H = 6–6.5; sp. gr. = 2.44–2.65. Color, white, gray, frequently flesh-red, salmon-red, greenish-white, green. Lustre vitreous, inclining to pearly on the cleavage planes.

Orthoclase is a silicate of alumina and potash, with the oxygen ratio of RO : Al_2O_3 : SiO_2 = 1 : 3 : 12. Some varieties contain a considerable quantity of soda, replacing potash.

It has been found in beautiful crystals, mostly twins, as well as in cleavable masses, throughout the gneissic rocks of the southeastern part of Pennsylvania, especially in the granitic segregations.

Good crystals have been found near Philadelphia, at Hestonville, forming a porphyritic gneiss; at the Columbia Bridge, and opposite Fairmount Water Works, the Flat Rock Tunnel on the Philadelphia and Reading Railroad, and McKinney's quarry in Germantown.

Very fine and large crystals some weighing 100 lbs., have been met with in Delaware county, especially in Lower Providence township, at George Sharples and Shoemaker's quarries; in Concord township; at Deshong's and Leiper's quarries in Ridley township; at Shaw and Ezra's quarry, near Chester, and Samuel Crozer's quarry, near Upland; very fine salmon-colored crystals are found in Henderson's quarry, near Upland; also in Birmingham and Edgemont townships.

Perfect crystals of orthoclase from five to six inches in size have been found in the muscovite at Swain's quarry in Pennsbury township; cleavable masses of white orthoclase occur near Unionville in Newlin township; and a blueish and grayish-white cleavable orthoclase, which emits a fetid odor, when struck, so-called "necronite" in the granular limestone of Caleb S. Cope's quarries, in East Bradford township; in Nivin's quarries in New Garden township; and Mendenhall's quarry in Pennsbury township; the variety "chesterlite" is found at the Poorhouse quarries, West Bradford township; and Baily's quarries, East Marlborough township, Chester county.

A fine green semi-transparent variety, almost without cleavage, the "lennilite" is found near Lenni, in Delaware county.

A white or grayish, sometimes inclining to a pale purplish, variety with definite cleavage, and a pearly, sometimes satin-like lustre, the "delawarite," occurs together with the last named between Lenni and Glen Riddle, Delaware county, and near West Chester, Chester county.

The more laminate and glassy semi-transparent blueish-green feldspar, the "cassinite," is found at Blue Hill, in Upper Providence township, Delaware county.

Green orthoclase, sometimes in very bright green cleavage-masses, but more usually pale green and much mixed with brownish tints, is found at Mineral Hill, Middletown, and in Upper Providence township, Delaware county.

Orthoclase having goethite disseminated through its mass and approaching *sunstone* in appearance occurs at Frankford.

Orthoclase in imperfect minute flesh-colored crystals (similar to those accompanying native copper of Lake Superior), has been found with garnet and hornblende, at St. Mary's near Knauertown, Warwick township; also with melanite in Warren township, Chester county.

A beautiful variety of orthoclase occurs at Van Arsdale's quarry, near Feisterville, Bucks county. It is sometimes found in perfect crystals of from a half to two inches in size, but mostly in cleavable masses of gray or grayish-black colors; sometimes it is opalescent with rich blue colors, much resembling the best varieties of labradorite.

Orthoclase has also been observed one mile east of Reading.

I have mentioned above that I had noticed striations in a feldspar from Seissholzville, near Alburtis, which had occurred in a rock composed of feldspar and the pyroxene variety, called *sahlite*.

The greater part of this feldspar does not show very distinct cleavage, and *no striations*, and the analysis, made at my request by Dr. W. P. Headden, proves it to be orthoclase.

The only orthoclase varieties from Pennsylvania, which have been analyzed are this, and the variety "chesterlite," (sp. gr. = 2.513. B. Silliman, Jr.) by Smith and Brush (Am. Jour. Sc., [2] XVI, 42):

		Chesterlite. Smith and Brush.		Seissholzville.* Headden.
		1.	2.	3.
Silicic acid	=	64.76	65.17	66.86
Alumina	=	17.60	17.70	18.97
Ferric oxide	=	0.50	0.50	0.62
Magnesia	=	0.30	0.25	trace
Lime	=	0.65	0.56	1.41
Soda	=	1.75	1.64	3.61
Potash	=	14.18	13.86	10.01
Ignition	=	0.65	0.65	——
		100.39	100.33	101.51

* The material was as pure as could be selected from the rock mixture; the analysis was only intended to establish the species to which the feldspar belongs.

The feldspars have undergone in some localities considerable alterations, and have given rise to kaolin, which is extensively mined in some parts of the State.

The unaltered cleavable orthoclase is also extensively mined in some parts of Delaware county, especially in Concord township; it is used for glazing porcelain, for the manufacture of artificial teeth, etc.

---

3. *Subsilicates.*

80. *Chondrodite* (*d'Ohsson*).

Occurs in orthorhombic crystals, often highly modified; in Pennsylvania only in crystalline grains. H = 6; sp. gr. = 3.1–3.24; usually between yellow and brown.

The chemical composition is a silicate of magnesia, in which part of the oxygen is replaced by fluorine; a portion of the magnesia is generally replaced by some ferrous oxide.

Near Brinton's Ford, on the Brandywine, Pennsbury township, Chester county, abundantly in orange and yellow grains, disseminated through the limestone.

It has not been analyzed.

81. *Tourmaline* (*Garmann*).

Rhombohedral crystals usually with unlike development at the two extremities; the prisms often triangular. Crystals frequently longitudinally striated; also in columnar, fibrous and in radiating masses, sometimes compact. H = 7–7.5; sp. gr. = 2.94–3.3. Colors usually black, brownish-black, also brown; green, blue, red to colorless. Lustre, vitreous. The chemical composition is very variable, and not fully understood, owing to some doubts about the state of the oxydation of the iron and the position of the boric acid in the constitution of the mineral. Most of the varieties are combinations of silicic and boric acids with alumina, ferric oxide, magnesia, ferrous oxide, small quantities of alcalies and a part of the oxygen replaced by fluorine.

Prof. Rammelsberg (Pogg. Annalen LXXX, 409 ff.) publishes in his investigations on tourmaline the analysis of a black tourmaline from Texas, Pennsylvania; but it is doubtful if this mineral occurs at this locality; however, as it is the only analysis of a Pennsylvania tourmaline which we possess, I will give the result:

Sp. gr. = 3.043. Loss by ignition = 3.30.

Silicic acid	=	38.47
Boric acid	=	8.48
Alumina	=	34.56
Ferric oxide	=	3.31
Manganous oxide	=	0.09
Magnesia	=	9.11
Lime	=	0.71
Soda	=	2.00
Potash	=	0.73
Phosphoric acid	=	0.20
Fluorine	=	2.36
		100.00

Beautiful crystals of bright yellow, also brownish yellow, and rarely white tourmaline are found in Bailey's limestone quarry, East Marlborough township; yellow tourmaline at Logan's limestone quarry, West Marlborough township; light yellow and sometimes transparent, also brown tourmaline in John Nivin's limestone quarry, New Garden township; brown tourmaline in beautiful hexagonal crystals in the limestone quarries of Wm. Jackson and Michiner, in London Grove township; and in Kennett township, Chester county.

Blueish and brownish-green tourmaline is found in fine crystals, penetrating damourite and diaspore at Unionville, Newlin township.

The minute tourmaline crystals, in entering from a cryptocrystalline white damourite into fine scaly granular prochlorite, are sometimes altered into the latter mineral, retaining their tourmaline form.

Green tourmaline is found in talc near Rock Springs, Lancaster county.

It is also met with in Marple township; a pale green and red columnar variety occurs rarely in albite at Leiperville, Delaware county.

Very fine prismatic crystals of black tourmaline, from one and a half to four inches in length and one to one and three-fourths in thickness, with perfect rhombohedral terminations, occur in the orthoclase quarry near Unionville, Newlin township;

these crystals are frequently broken and the fragments cemented by granular quartz or orthoclase; smaller but equally fine black tourmaline crystals, terminated by planes of two obtuse rhombohedra, are found at Lewisville, Elk township; black tourmaline in prismatic crystals in quartz is found near Oxford; and velvet black prisms with rhombohedral terminations at Wm. Jackson's farm, London Grove township; it occurs in cylindrical crystals in quartz on Jos. Osborne's farm, Westtown township; at Painter's farm, East Bradford township; in New London township; and at Parkesburg, Sadsbury township; acicular crystals of black tourmaline occur in white quartz near the Marlborough Meeting-House, East Marlborough; and in muscovite in Thornbury township; in finely crystalline columnar masses it is found in mica schist near Amos Pugh's mill, East Nottingham township, Chester county.

Very beautiful crystals of black tourmaline are found in Delaware county; near Leiperville, for instance, it is found in crystals of five inches in length and one and one-half inches thick, and well terminated; also in Marple township, terminated with two flat rhombohedra; good specimens of crystallized black tourmaline have also been found on Wm. Eyre's farm and at Trainer's Mill-dam, near Upland, and in Aston, Upper Providence and Springfield townships.

Radiated, stellated and fibrous black tourmaline is found at Rockdale, Upper Providence and Middletown townships; also occasionally, together with smaller crystals, in a schistose rock at Deshong's quarry, in Ridley township, Delaware county.

Near Philadelphia, black tourmaline has been found in the gneissic rocks at Frankford; in fine black crystals and radiated aggregations at West Washington Lane, Gorgas Lane, McKinney's quarry, &c., near Germantown; with beryl on York Road, five miles from Philadelphia; on the Falls of the Schuylkill and opposite Fairmount Water-works.

In Bucks county tourmaline has been found in prismatic, longitudinally striated crystals with rhombohedral termination in granitic veins in gneiss, at Nevil's Academy, near Bustleton.

Black tourmaline is found near Easton in quartz, and five miles east of Bethlehem in fibrous masses and imperfect crystals, associated with allanite.

82. *Andalusite* (*Delamétherie*).

In rhombic prisms with an angle of 90° 48′, and therefore much resembling quadratic prisms; some of the crystals slightly modified. The crystals, which have been found in Pennsylvania are usually very rough, rarely with smooth planes. E. S. Dana (Am. Jour. Sc. [3] IV, 473) has described a remarkable crystal from Upper Providence, Delaware county. H. = 7.5; sp. gr. = 3.15. Colors grayish and reddish-white. Lustre, vitreous.

In its pure state, silicate of alumina, Al_2O_3,SiO_2 = silicic acid, 36.8, alumina, 63.2.

Delaware county has furnished some of the largest single crystals and groups of crystals, known in the world.

On the farm of James Worrell, in Upper Providence township, a crystal of seven and a-half pounds in weight has been found, and a group of crystals of about sixty pounds in weight; good specimens of finely modified crystals have been found in Springfield township, one-half mile north of Westdale; very good crystals, sometimes terminated at both ends, have been found in Marple township. It also occurs in Lower Providence township.

83. *Fibrolite* (*Bournon*).

This very interesting mineral occurs in Pennsylvania, at several localities, under circumstances which show that it has played an important part in the alteration of other minerals, being both the result of the change, and the material which has been converted into other species.

It is found in monoclinic crystals, with eminent orthodiagonal cleavage and subadamantine lustre upon the cleavage planes. In Pennsylvania *no* crystals have been found but only fine and coarse fibrous, sometimes columnar masses, occasionally radiating. H. sp. gr. = 3.2–3.3. Colors white, grayish, yellowish and brownish-white, gray.

Its chemical composition is identical with that of andalusite.

In the neighborhood of Philadelphia, in Germantown, it occurs in very fine fibres in quartz, with cyanite in the centre; also abundantly on the Wissahickon, at Columbia Bridge and elsewhere in West Philadelphia.

In Delaware county it is found in fine fibrous masses, at Bullock's quarry, Birmingham township; in feldspar, near the Black Horse, in Middletown township; and half way between this place and Village Green, Aston township; at Village Green in small fibres in quartz; in Concord township; also at the White Horse, the celebrated cyanite locality, from three to four miles below Darby, in Ridley township; and at Dutton's mill, near Chester, it occurs in fibrous masses.

Little bundles of delicate fibres and acicular crystals of a grayish-white color are found in Nivin's limestone quarry, New Garden township, Chester county.

Very interesting are congeries of fibrous and radiating, microscopic crystals, in the form of corundum, resulting from its alteration, some with a nucleus of unaltered corundum, (see my paper on corundum, *l. c*); they occur in a feldspathic rock near Media, Middletown township.

A mineral which has been found in the broken stone on the Delaware county turnpike, near Abram Powell's dwelling, in Darby township, and which generally has been mistaken for zoisite, is probably fibrolite, pseudomorphous after andalusite. The crystals which I have seen are short and stout, but too imperfect for me to speak on this point with certainty; it is from pale to dark ash-gray, shows a perfect cleavage with strong subadamantine lustre and, as my analysis shows, has the composition of fibrolite.

Mr. E. S. Dana (*l. c.*) states, in his description of the Upper Providence andalusite, that the cleavage of this mineral from Delaware county was in most cases irregular, many of the crystals having a fibrous, tremolitic and in others a radiated structure. This is in all probability owing to a paramorphism, an alteration of andalusite into fibrolite, which I have observed on specimens from North Conway, N. H., in which the original andalusite first changes into a mineral, like fibrolite or cyanite, before it is altered into scaly damourite.

I will mention in this connection the peculiar fibrous and radiating minerals, which Dr. Isaac Lea (Proc. Ac. Nat. Sc. Phil. 1867, 44), called "Lesleyite."

As I showed in my investigations on corundum (*l. c.*), these fibrous lesleyites probably result from the alteration of corundum, and are fibrolites in different stages of alteration into damourite and other species.

Fibrolite from several localities of Pennsylvania has been analyzed, and the result of some of the analyses agreeing with that of cyanite, Lardner Vanuxem (Journ. Ac. Nat. Sc. Phila. VI, 1. 41) proposed to unite it with the latter specie under the name "disthene."

The analysis of the fibrolite from Chester, Delaware county, was made first by Thos. Thomson (Ann. Lyc. N. H. New York, III, 41) and subsequently by B. Silliman, Jr. (Am. Journ. Sc. [2] VIII, 388), it was a third time analyzed by A. Erdmann (Ak. H. Stockholm, 1842, 19), (1) a. b. c.; the fibrolite resulting from the alteration of corundum from Media, was analyzed by myself (2); that which is probably an altered andalusite from Darby township, by myself (3); the least altered "lesleyite" from Unionville, by Dr. G. A. Koenig and myself (my paper on Corundum, *l. c.*), (4) a. and b.

	1. Fibrolite from Chester, Delaware county.			2. Pseudomorphous, from Media. GENTH.*	3. Pseudomorphous, from Darby, GENTH.	4. Lesleyite, least altered.	
	a. THOMSON.	*b.* SILLIMAN.	*c.* ERDMANN.			*a.* KŒNIG.	*b.* GENTH.
Sp. gr. =	——	——	3.239	3.286	——	3.157	——
Silicic acid =	46.40	35.96	40.05	37.37	36.98	34.80	35.68
Alumina =	52.92	64.43	58.88	[60.52]	62.85	59.77	60.29
Ferric oxide =	trace	——	0.74	0.90	trace	0.73	0.72
Manganous oxide =	——	——	——	0.10	——	——	——
Magnesia =	——	0.52	——	0.25	——	——	0.29
Lime =	——	——	——	0.38	——	——	——
Lithia =	——	——	——	——	trace	——	trace
Soda =	——	——	——	——	0.24	not det.	0.41
Potash =	——	——	——	——	0.22	" "	0.96
Ignition =	——	——	0.40	0.48	0.66	2.05	1.78
Corundum =	——	——	——	——	——	2.20	——
	99.32	100.91	100.07	100.00	100.95	99.55	100.13

* I have made two other analyses of the same, but give only the results made with the best material.

84. *Cyanite* (*Werner*).

The crystals belong to the triclinic system, but distinct crystals are very rare, and have not been observed in this State. Usually in very long bladed crystals or crystalline aggregations, in blades from nearly one inch in width almost to sub-fibrous. H. unequal on the different planes from 5 on the lateral, to 7.25; sp. gr. = 3.5 to 3.7. Colorless and white, blueish-white to deep azure-blue; the deeper blue color usually in the centre of the crystals; also greenish-blue, gray and grayish-black. Lustre vitreous, on cleavage planes inclining to pearly. The composition identical with the two preceding species.

It is a very remarkable fact that the silicate of alumina, $Al_2O_3SiO_2$, occurs in three different systems of crystallization; in nature we find that at times crystals have the form of one species and the structure of another, as I pointed out to be probably the case with the fibrolite from Darby, which appears to be a pseudomorph or paramorph of andalusite, and the white fibrolite of Germantown which has sometimes blue cyanite in its centre, and some of the so-called "bucholzites," from the neighborhood of Philadelphia, which frequently have a close resemblance to cyanite, and appear to graduate into this species.

Very beautiful specimens of cyanite in several varieties have been found in Delaware county.

In blue, radiating, bladed masses in granular quartz, sometimes associated with beryl, pyrite and *fibrolite*, it is found at the White Horse, from three to four miles below Darby; and in Ward's quarry, near Leiperville in Ridley township; near the first locality is also found a gray or grayish-black cyanite in imperfect crystals both in decomposed mica schist, or loose in the soil; in the same county cyanite is found on Mt. Zion Hill, near Darby; and with staurolite in mica slate on Mrs. Pritchett's farm in Haverford township.

In the neighborhood of Philadelphia light to deep blue bladed crystalline cyanite is found in gneissic rocks in Allen's, Gorgas' and Crease's Lanes and on the Wissahickon.

Gray-bladed cyanite is found in mica slate and in detached crystals sometimes three inches in length and one inch broad near Cope's Mill, and blue cyanite occurs at Caleb Cope's limestone quarry, East Bradford township; also blueish-green bladed crystals in (?) albite, half way below the Black Horse and Cope's; green cyanite is found near Marshalton, West Bradford township; and a blueish variety one mile east of the Poorhouse; blue-bladed cyanite is found in granular quartz in Bailey's and near Logan's quarry in West Marlborough township; also in Wm. Jackson's quarry and near the Meeting House in London Grove township. Cyanite associated with damourite occurs near Unionville, Newlin township, one mile north of the corundum mine.

Very fine blue, flat crystals and blades are found in quartz at Nevil's Academy, near Bustleton, Bucks county.

None of the Pennsylvania cyanite varieties has been analyzed.

85. *Titanite* (*Klaproth*).

Crystallizes in monoclinic, often highly modified crystals. It is a rare mineral in Pennsylvania, and but few persons have had an opportunity to see good crystals, the forms of which have not yet been studied. Mostly in crystals. H = 5–5.5 ; sp. gr. = 3.4–3.56. Colors, brown, grayish-brown and yellow. Lustre, resinous to adamantine. A silico-titanate of lime = $CaO,2SiO_2+CaO,2TiO_2$.

Waxy yellow and also some fine brown crystals, occasionally over one inch in diameter have formerly been obtained in hornblendic gneiss on the Schuylkill, near Philadelphia; also on Schuylkill Falls; in small but perfect brown crystals at Frankford; at McKinney's quarry and at Heft's Mill.

It is found on Osborne's Hill, East Bradford township; in brown hemitrope crystals at David Nivin's limestone quarry, New Garden township; in brown imperfect crystals, sometimes about one inch in diameter, with green pyroxene, at Brinton's Ford; in small brown crystals about half an inch in diameter, and in yellow crystals with sunstone at W. Cloud's farm and Pearce's paper-mill, in Kennett township, Chester county.

It is found in large imperfect, crystals, implanted in allanite or brown garnet, sometimes much decomposed, five miles east of Bethlehem; and in small imperfect brown crystals in limestone, three miles north of Bethlehem; also in the neighborhood of

Easton, and in good brown crystals in Van Arsdale's quarry, Bucks county.

It has also been observed in Chrisman's limestone quarry, South Coventry, Chester county; and two miles northeast from Jones' Mine, Berks county.

86. *Staurolite* (*Delamétherie*).

In orthorhombic prisms, frequently in twins, forming cross-crystals; the crystals have often rough surfaces; H. = 7–7.5; sp. gr. = 3.4–3.8. Color usually dark reddish-brown to brownish-black. Lustre subvitreous, inclining to resinous. Its chemical composition is not fully established; principally silicate of alumina and ferric oxide with ferrous oxide and magnesia.

Near Philadelphia it is found at Prince's soapstone quarries in chlorite schist in small brown crystals, which sometimes appear broken and pulled apart lengthwise; it is also met with at the Columbia Bridge; abundant at Megargee's paper-mill, on the Wissahickon, and at Gorgas' and Crease's Lanes; also in mica slate at Glen Riddle Station, Twenty-fourth Ward, Philadelphia.

In Delaware county it occurs withcyanite in mica slate on Mrs. Pritchett's farm, and with garnets east of Friends' Meeting-house, Haverford township; and loose in the soil near Athensville in large rough crystals; also near Tyson's Mill, and elsewhere in Aston township; at Beattie's Mill, Springfield township, and in Darby and Middletown townships.

Staurolite has undergone a remarkable change in the well known steatite bed between Chestnut Hill and Mill Creek, and has been altered into serpentine, as observed by Mr. Th. D. Rand (Proc. Ac. Nat. Sc., Phila., 1872, 303).

b. *Hydrous Silicates.*

Arrangement of species:

a. The General Section of Hydrous Silicates.

It includes all hydrous silicates, excepting the *Zeolites* and *Margarophyllites*.

1. Bisilicates.
2. Unisilicates.
3. Subsilicates.

β. Zeolite Section.

Mostly hydrous silicates of alumina with alcaline earths and alcalies, in constitution much resembling the feldspar-group; the oxygen ratio of protoxides and sesquioxides being 1 : 3.

γ. Margarophyllite Section.

Micaceous or thin foliated, when crystallized ; the plane angle of base of prism 120°. Very variable in their chemical constitution, being:

1. Bisilicates.
2. Unisilicates.
3. Subsilicates.

---

a. General Section of Hydrous Silicates

1. *Bisilicates.*

87. *Laumontite* (*Haüy*).

In small monoclinic crystals with perfect prismatic and clinodiagonal cleavages. H. = 3.5–4; sp. gr. = 2.25–2.36. White, passing into yellowish and grayish white. Lustre vitreous. A hydrous silicate of alumina and lime, with the oxygen ratio for $CaO : Al_2O_3 : SiO_2 : H_2O = 1 : 3 : 8 : 4$.

The crystals usually loose water and fall to powder; but the specimens found at Columbia Bridge keep remarkably well; it also occurs at the Falls of the Schuylkill. It has also been found, colored green by copper, on the Wissahickon, two miles from its mouth, and in small white crystals at McKinney's quarry, near Germantown.

88. *Chrysocolla* (*Haidinger*).

An amorphous or crypto-crystalline mineral, often in mammillary masses and concretions, sometimes botryoidal and stalactitic; in incrustations or filling seams. H. = 2–4; sp. gr. = 2 –2.24. Color between blue and green, from sky- and turquois blue to mountain-green; also brownish and black. Lustre vitreous, waxy and dull. The composition is very variable, owing to frequent admixtures of other minerals, such as allophane, malachite, &c.

Beautiful massive specimens, also botryoidal or stalactitic coatings and globular aggregations of a greenish-blue color, have been found at Jones' Mine, near Morgantown, Berks county.

In similar specimens of great beauty, of various shades between blue and green, sometimes of a blackish-green color, it occurs at the Cornwall Mines, Lebanon county, where it has also been met with in pseudomorphs after dolomite.

It is also found at the Elizabeth Mine, in Warwick township, and at the mines near Phœnixville, in Chester county; at the Perkiomen Mines, Montgomery county, and sparingly at Frankford and on the Wissahickon.

If it would occur in larger masses it would be a valuable copper ore, containing in its purer varieties from twenty to forty per cent. of copper.

None of the Pennsylvania varieties has been analyzed.

89. *Neolite* (?) (*Scheerer*).

A massive yellowish or grayish-white, amorphous mineral of a waxy lustre, filling fissures in serpentine rocks and probably resulting, like this, from the alteration of pyroxenic minerals or bronzite, occurs at Rose's serpentine quarry, opposite Lafayette, in Montgomery county; it may belong to this species.

Mr. Harry W. Jayne has, in the Laboratory of the University of Pennsylvania, at my request, made an analysis of it, and found:

Silicic acid	=	61.70
Alumina, Ferric oxide	=	6.85
Magnesia	=	27.95
Water	=	5.00
		101.50

2. *Unisilicates.*

90. *Calamine* (*Smithson*).

Crystallizes in hemimorphic-hemihedral, orthorhombic forms; also stalactitic, mammillated, botryoidal, fibrous and granular massive. H. = 4.5–5; sp. gr. = 3.16–3.9. White, into yellowish and brownish, also greenish-blue. Lustre vitreous. A hydrous silicate of zinc, usually $2ZnO, SiO_2+H_2O$, containing: Silicic acid, 25.0, Zincoxide, 67.5, Water, 7.5.

A very valuable zinc ore, which is largely mined at Friedensville, Lehigh county. Here it occurs, sometimes upon aragonite, in very beautiful, although small, crystals, and groups of crystals; also granular massive.

At Espy, near Selin's Grove, in Northumberland county, it occurs in small quantity with zincblende.

In snow-white silky tufts and yellowish or blueish crystalline masses and minute crystals, it has been found at the Wheatley Mines, near Phœnixville.

Small crystals, also globular and botryoidal crystalline aggregations, often of a greenish-blue color, from a small percentage of copper, were found at the Ecton Mine, Montgomery county.

In Blair county, in Sinking Valley, it has been sparingly met with.

In minute crystals it occurs at the Lancaster zinc mines, five miles south of, and four miles northwest of Lancaster, on the Pennsylvania Railroad.

91. *Prehnite* (*Werner*).

It is reported as occurring at Marsh Creek, in the neighborhood of Gettysburg, Adams county

I have not seen any from this State.

92. *Apophyllite* (*Haüy*).

Crystallizes in forms belonging to the tetragonal system, with highly perfect cleavage parallel to the basal plane. H. =4.5–5; sp. gr. = 2.3–2.4. Colorless, white. Pearly lustre upon the cleavage plane, otherwise vitreous. A hydrous silicate of lime and potash, a small portion of the oxygen being replaced by fluorine. $K_2O,2SiO_2+8(CaO,2SiO_2)+18H_2O$.

This beautiful mineral had been observed by Mr. Th. D. Rand, in a narrow fissure in the gneiss of Frankford, in crystalline masses, showing rarely octahedral and basal planes; I have seen a similar specimen from McKinney's quarry.

Last summer Mr. H. W. Hollenbush, of Reading, discovered a highly interesting locality at Fritz's Island, near Reading, where it is found, associated with calcite and several zeolites in a granular garnet rock. To him and Mr. Samuel Tyson, of King of Prussia, Montgomery county, I am indebted for a liberal supply of specimens.

Usually it occurs in tetragonal tabular crystals, rarely without any other planes, mostly with octahedral planes (the form 387, in Dana's Mineralogy). These crystals sometimes form twins or are in groups of rosettes. On one specimen I have observed microscopic, but beautiful and perfect octahedral crystals, either with small planes of the second prism or even without modification.

It has not been analyzed.

3. *Subsilicates.*

93. *Allophane* (*Stromeyer*).

Amorphous. Usually in incrustations with mammillary surface, and stalactitic. H. = 3; sp. gr. = 1.85–1.89. Colorless, white, blueish, sky-blue. Lustre vitreous to resinous, sometimes waxy and pearly. Mostly a hydrous silicate of alumina with the oxygen ratio of $Al_2O_3 : SiO_2 : H_2O = 3:2:6 = Al_2O_3,SiO_2 + 6\,H_2O$.

It is found in very fine white and sky-blue mammillary and stalactitic masses at Cornwall, Lebanon county; under similar circumstances it occurs at Jones' Mine, near Morgantown, Berks county.

At the Friedensville Zinc Mines, near Bethlehem, it is found in white botryoidal and stalactitic masses.

β. *Zeolite section.*

94. *Thomsonite* (*Brooke*).

Crystallizes in forms belonging to the orthorhombic system. Also columnar, radiated; in radiated spherical concretions and compact. H. = 5–5.5; sp. gr. = 2.3–2.4. White, brownish-white. Lustre vitreous, inclining to pearly and waxy. A hydrous silicate of alumina, lime and soda.

Very small spherical concretions of very fine radiated structure, whitish color and a waxy lustre, inclining to pearly occur in small quantities with other zeolites, at Fritz's Island, near Reading.

With great difficulty I selected about two and a half grains of tolerably pure material, which by analysis gave me results, leaving very little doubt that the mineral is thomsonite.

I found:

Silicic acid	=	38.74
Alumina	=	[31.22]
Lime	=	10.77
Soda	=	3.32
Potash	=	0.45
Water	=	15.50
		100.00

It is probable that some other of the zeolites from Pennsylvania, which I have seen in collections, may be thomsonite.

95. *Mesolite* (*Fuchs and Gehlen*).

Also at Fritz's Island occurs a zeolite, either in very minute white tufts or radiating white needles, or in globular concretions of very fine white fibres. It is so rare that no attempt could be made to examine it, but its appearance and association make it probable that it is mesolite, another species of a hydrous silicate of alumina, lime and soda.

96. *Chabazite* (*Bosc d'Antic*).

In rhombohedral crystals with an angle of 94° 46′; sometimes modified by other rhombohedra and scalenohedra. H = 4–5; sp. gr. = 2.08–2.19. White, yellow, brown, reddish.

Lustre, vitreous. Mostly a hydrous silicate of alumina and lime, with the oxygen ratio of $RO : Al_2O_3 : SiO_2 : H_2O = 1 : 3 : 8 : 6$, corresponding to $CaO, SiO_2+Al_2O_3,3SiO_2+6H_2O$; a small quantity of lime is sometimes replaced by soda and potash.

Beautiful colorless crystals are found at Fritz's Island, rarely showing truncation of edges by a second rhombohedron.

In small yellow rhombohedra it is found at Schuylkill Falls; in brown rhombohedra at Flat Rock Tunnel on the Philadelphia and Reading Railroad; in red crystals at Turtle Rock in Fairmount Park near Philadelphia; in reddish-brown rhombohedra it occurs on Robert Lamborn's farm, West Bradford township, and in reddish rhombohedra with sunstone at Cloud's Wall, Kennett township, Chester county.

None of the Pennsylvania varieties have been analyzed.

97. *Stilbite* (*Haüy*).

It is found in modified orthorhombic crystals with perfect brachydiagonal cleavage; usually in sheaf-like aggregations, divergent or radiated. It has a very characteristic pearly lustre upon the cleavage planes, on the others a vitreous lustre.

$H = 3.5–4$; sp. gr. $= 2–2.2$. Color white, yellow, brown.

This is another hydrous silicate of alumina and lime (with small quantities or soda and potash) having an oxygen ratio of $RO : Al_2O_3 : SiO_2 : H_2O = 1 : 3 : 12 : 6$, corresponding with: $CaO,3SiO_2+Al_2O_3,3SiO_2 + 6H_2O$.

In Pennsylvania it has been found near Philadelphia in fissures of gneissic rocks at Flat Rock Tunnel, at McKinney's quarry and Wayne Station; in white imperfect crystals and radiated masses in hornblendic rocks on the Schuylkill, four miles from the city, and in beautiful white radiated coatings and masses, rarely one inch in thickness at Frankford.

In yellowish-brown radiated columnar masses it occurs on quartz in gneiss at Leiperville, Delaware county.

In white fascicular groups of minute crystals and in radiating masses of pearly lustre in narrow veins in hornblende rocks on Rob't Lamborn's farm, West Bradford township, Chester county. It also has been found at Rautenbush, near Reading.

No Pennsylvania variety has been analyzed.

98. *Heulandite (Brooke).*

In monoclinic crystals with eminent clinodiagonal cleavage, rarely in globular and radiating forms. H = 3.5–4; sp. gr. = 2.2. White, yellowish, brown. Lustre vitreous, on the cleavage planes strongly pearly. Its composition is very nearly the same as that of stilbite, perhaps with one atom less water. It occurs, associated with stilbite, in small brownish crystals at Columbia Bridge, McKinney's quarry, and Flat Rock Tunnel, near Philadelphia.

99. *Undescribed Zeolite.*

Associated with apophyllite, chabazite, &c., occurs at Fritz's Island a zeolite, which appears to be new.

It has not yet been found in crystals sufficiently distinct and large to determine their form. The crystalline groups which have been observed, much resemble rectangular prisms with rhombic octahedra, twinned like Phillipsite and surrounding a nucleus of calcite. Frequently the latter is weathered out, leaving a shell of the zeolite.

It has been impossible to obtain for analysis material, which was free from inclosed calcite; the best, which I was able to select, was found to contain 9.75 per cent. of the same, which was deducted from the analysis, giving for the pure zeolite the following composition:

Silicic acid	=	43.36	contains oxygen,			22.51
Alumina	=	28.78	"	"		13.44
Lime	=	10.95	"	"	3.11	
Soda	=	0.68	"	"	0.18	3.52
Potash	=	1.38	"	"	0.23	
Water	=	15.52	"	"		13.80
		100.67				

The oxygen ratio of RO (including R_2O,): Al_2O_3 : SiO_2 : H_2O = 3.52 : 13.44 : 22.51 : 13.80 = 1 : 3.82 : 6.39 : 3.92.

The oxygen ratio of Levynite is nearly the same, viz., 1 : 3 : 6 : 4; but their forms are evidently different.

It requires better material of this interesting zeolite, before its true character can be established by a fuller crystallographic and chemical investigation.

Crystallized zeolites have been observed in several varieties in Adams county and elsewhere in Pennsylvania, but they have not been distinguished.

I have not seen any from other localities than those above mentioned.

γ. *Margarophyllite Section.*

1. *Bisilicates.*

100. *Talc* (*Agricola*), *Steatite* (*Cronstedt*).

Orthorhombic, but rarely in distinct hexagonal prisms or plates. Foliated massive, sometimes in globular or stellated groups; also coarse or fine granular, crypto-crystalline or compact. Cleavage basal eminent. H = 1–1.5; sp. gr. = 2.55–2.8. Colors white, silvery-white, greenish-white, grayish-white; some of the compact varieties greenish-gray. Lustre, pearly.

Talc and steatite are a hydrous silicate of magnesia, of variable composition, some varieties with, others without water.

Large masses of green, white and silvery-white talc, in small folia, together with the compact crypto-crystalline variety (steatite) occur at Easton.

Laminated, sometimes of a rich green, usually, however, of a greenish or grayish-white color, also fibrous and compact, it occurs from Chestnut Hill to Spring Mill, and is largely mined at the soapstone quarries near Lafayette, Montgomery county.

Fibrous and scaly talc are mentioned by C. F. Schaeffer (Journ. Ac. N. Sc., Phila., 1819, 236) as occurring in granite at Roxborough, near Philadelphia; and radiated steatite in tufts of acicular crystals, and of silky lustre, by Jacob Porter, June 20, 1824 (Journ. Ac. Nat. Sc., Phila., VIII, 233), as being found on the Wissahickon, one mile above its mouth.

In Delaware county talc in various varieties, has been met with; abundantly in the serpentine and the subjacent strata, for instance, in Radnor, Marple, Middletown, and Aston townships.

The whole serpentine range of Chester and Lancaster counties is rich in localities, where talc has been found.

In the former, it is found in West Goshen township, three

miles south of West Chester, both fine scaly and steatitic; of a pale greenish-gray color it occurs near Unionville, Newlin township; also steatitic and white foliated and in scales in Michiner's quarry, London Grove township; in foliated masses of a greenish-white color, near the chrome mines, in West Nottingham and in East Fallowfield townships.

In Lancaster county it is found in foliated greenish-white masses near the Texas Chrome Mines; at Rock Springs; Reynold's Mine; Low's Mine, &c.

Talc has been met with at Jones' Mine, Berks county, and ten miles south of Carlisle, in the South Mountains.

A peculiar mineral, of a dull greenish-gray color, and apparently amorphous, forms seams of sometimes a thickness of two inches, at the Black Horse, East Bradford township, Chester county, where a spur of serpentine rests upon hornblendic rocks.

It is probably the result of the decomposition of the hornblende rocks, and looks as if it had been a mud, which dried rapidly, and is full of cracks from shrinkage. It is usually called "indurated talc," but is probably *not* talc, but something else.

Similar varieties occur in West Goshen township, Chester county, and in Middletown township, Delaware county.

Neither this nor any other of the Pennsylvania varieties of talc have been analyzed.

101. *Pyrophyllite* (?) (*Hermann*).

Agalmatolite is mentioned as occurring in Penn township, Chester county.

I have never seen it, and therefore am not prepared to state whether or not it is the compact variety of pyrophyllite.

102. *Sepiolite* (*Glocker*).

Sepiolite or meerschaum has occasionally been met with in compact masses of a smooth earthy texture in the magnesia quarries in West Nottingham township, Chester county. Only a few small pieces have been found, but they were of good quality.

It also occurs in grayish and yellowish-white masses in the

serpentine near Stamp's Tavern, in Concord township, Delaware county.

Neither has been analyzed.

2. *Unisilicates.*

103. *Serpentine* (*Wallerius*).

Serpentine being the product of alteration of other minerals and rock masses has therefore no form of its own, but frequently retains that, which belongs to the original.

As many mineral species have been changed into serpentine, we necessarily must expect it to occur in many pseudomorphous forms.

Much has been said and written on this subject. Lately a most beautiful investigation on Serpentine Pseudomorphs, and others, from the Tilly Foster Mine, N. Y., has been published by Prof. J. D. Dana (Am. J. Sc. [3] VIII, 371–381 and 447–459). Although not immediately connected with Pennsylvania Mineralogy, I did not think it out of place to refer to this important paper, partly because some of the alterations therein mentioned, may occur in this State, but principally because a more extended study of our own serpentine minerals may throw some light upon many yet unsettled points.

In a paper, which I published many years ago (Am. Journ. of Sc. [2] XXXIII, 199), I showed reasons for a suggestion, which I then made, that our whole chromiferous serpentine range results from the alteration of chrysolite. An opinion which since has been repeatedly corroborated by investigations made in many other parts of the world.

Lenticular beds of serpentine occur in many localities in the southeastern portion of Pennsylvania, within the gneissic rocks, frequently with a bed of bronzite rock, containing sometimes grains of chrysolite, in the immediate neighborhood. As the exact position of these rocks has not been made out, I refer to them as a subject, which requires much fuller investigation.

Some of the massive varieties of serpentine, like that from Rose's Quarry, opposite Lafayette, Montgomery county, and especially the green varieties from Brinton's Quarry, and the grayish-green from West Nottingham, Chester county; and Crump's, near

Media, Delaware county, have been used for many years as building stones, and have lately become quite fashionable in this city; the University of Pennsylvania buildings, many churches and other prominent edifices, having been erected of the same.

Serpentine is an amorphous mineral, usually of different shades of green, yellowish oil-green, siskin-green, leek-green, blackish-green, gray, yellowish-brown, sometimes white or greenish or yellowish-white. Usually massive, also slaty and pseudomorphous.

H. = 2.5–4; sp. gr. = 2.2-2.65. Lustre sometimes that of the original mineral; for instance, silky in some fibrous varieties; usually from subresinous to greasy, resinous, waxy; also dull.

The composition of serpentine is mostly a combination of silicate of magnesia, with magnesia hydrate, the oxygen ratio of $MgO : SiO_2 : H_2O$: being = 3 : 4 : 2, corresponding with : $2MgO,SiO_2 + MgO,2H_2O$.

a. Precious or *noble* serpentine of rich oil-green color has been found massive and in grains disseminated through calcite, and in fine pseudomorphous crystals after pyroxene, at Easton; it is also found in East Goshen township, Chester county. The Easton variety has been analyzed by Thos. Thomson (Ann. Lyc. N. Hist. New York, III, 49) and the latter by S. P. Sharples (Am. Jour. Sc. [2] XLII, 272). (See below.)

Precious serpentine has been found together with common, massive serpentine at Fritz's Island, near Reading, Berks county; and with common serpentine and several other varieties near the Yellow Springs Road, in Radnor township, Delaware county.

The massive *common* serpentine of a dark green color, often with white narrow veins of carbonate of lime and chrysotile, is found at Easton.

Very beautiful striped and mottled varieties of common serpentine of pale grayish or yellowish-green and a dark greenish-black color and a waxy lustre are found in West Nottingham township, Chester county. They would be very pretty for small ornaments, but are rather brittle.

Greenish-black serpentine, pseudomorphous after staurolite occurs abundantly, near Lafayette, Montgomery county, and in the same rocks northeast and southwest of it, between Chestnut Hill and Spring Mill. These pseudomorphous crystals, inclosed in the grayish-white soapstone, form a very beautiful rock.

Outside of the serpentine range, common serpentine has been observed in peculiar varieties which deserve fuller investigation: in Potsdam sandstone, a short distance north of Bethlehem; in the iron mines at Cornwall, Lebanon county, and near Knauertown, Warwick township, Chester county.

β. A *porcellanous* variety, resembling compact lithomarge, with smooth porcelain-like fracture, H. = 3.5 and sp. gr. = 2.48, occurs at Middletown, Delaware county. It has been analyzed by B. S. Burton (Dana's Mineralogy 1868, 467).

γ. The variety "*bowenite,*" of a fine granular texture and of a greenish or reddish-white color and great tenacity is found at Easton. It frequently contains a small quantity of tremolite intermixed. It has not been analyzed.

δ. A dark-green lamellar variety, somewhat resembling antigorite, occurs as a vein of from three to six inches in width in serpentine at Unionville, Newlin township, Chester county.

ε. A beautiful, sometimes laminated, compact mineral of H = 4.5 ; sp. gr. = 2.59–2.64, and various shades of green, from a pale apple-green to deep green, has been called *williamsite* by Shepard (Am. Journ. Sc. [2] VI, 249); G. J. Brush (Dana's Mineralogy 1850, 692) and subsequently Hermann (Journ. Prakt. Chem. LIII, 31) pointed out its identity with serpentine, which was again confirmed by additional analyses, by Smith and Brush (Am. J. Sc. [2] XV, 211). See analyses below.

It graduates into a granular, frequently slaty, serpentine of greenish-gray color. The best specimens of bright color come from Low's Mine; in inferior pieces it has been found at the Red Pit and Wood's Mine, all in Lancaster county.

A slaty serpentine from Texas, Lancaster county, has been analyzed by B. Silliman, Jr. (Dana's Mineralogy, 1850, 692).

ζ. The thinly foliated serpentine "*marmolite,*" resulting in most cases from the alteration of brucite, is found at Wood's Mine, Lancaster county; at Scott's Chrome Mine and the Magnesia quarries in West Nottingham township, and in West Goshen township, Chester county; also in Radnor township, Delaware county.

η. The so-called *chrysotile* is found in narrow seams in common serpentine. It consists of very delicate fibres of an olive-green or greenish-white color and a submetallic, silky lustre.

It is found at Blue Hill, in Upper Providence township, in a seam in decomposed serpentine, the fibres are sometimes four inches long; in narrow seams of the thickness of a knife's blade to half an inch at Moro Phillips' Chrome Mine, in Marple township; also in Radnor township, Delaware county; in a seam of about half an inch in thickness it occurs in quartz in West Goshen township, Chester county.

ϑ. The variety *picrolite* occurs in dark-grayish or blueish-green, fibrous masses at Wood's Mine, Reynold's Mine and Low's Mine, in Lancaster county; also in Radnor, Delaware county.

The green variety from the Wood's Mine has been analyzed by Rammelsberg (Mineralchemie, 526) and Brewer (Dana's Mineralogy, 1850, 692).

κ. The variety *baltimorite*, often mixed with dolomite, and of a greenish or reddish-gray color, has been met with at Low's Mine, Wood's Mine and Rock Springs Mine, Lancaster county.

The following are the results of the analyses of the various varieties of Pennsylvania serpentines:

		α. Precious S.			ε. Williamsite. * Low's Mine				ε. Slaty Serpentine. Texas.
		Easton.	E. Goshen.	β. Porcellanous S. Middletown.					
		THOMSON.	SHARPLES.	BURTON.	BRUSH.	HERMANN.	SMITH AND BRUSH.		SILLIMAN.
Sp. Grav.	=	3.39	—	2.48	—	2.60	—	—	—
Silicic acid	=	41.55	43.89	44.08	45.02	44.50	41.60	42.60	44 58
Alumina	=	—	—	0.30	3.35	0.75	trace	trace	3.03
Niccolous oxide	=	—	—	—	—	0.90	0.50	0.40	—
Ferrous oxide	=	—	1.38	1.17	—	1.39	3.24	1.62	6.15
Ferric oxide	=	3.90	—	—	—	—	—	—	—
Lime	=	—	—	0.37	—	—	—	—	—
Magnesia	=	40.15	40.48	40.87	37.75	39.71	41.11	41.90	34.51
Water	=	13.70	13.45	13.70	13 01	12.75	12.70	12.70	12.38
		99.30	99.20	100.49	99.13	100.00	99.15	99.22	100.65

* I do not give the results of the Shepard's analysis of Williamsite, because they have not been corroborated by subsequent investigators; Brush's first analysis is given merely to show his priority over Hermann, in pointing out the identity of Williamsite with serpentine; it has since been replaced by others of himself and Smith.

ϑ. Picrolite, Wood's Mine.

		Rammelsberg.	Brewer.
Sp. gr.	=	2.557	——
Silicic acid	=	43.79	44.25
Alumina	=	——	4 90
Niccolous oxide	=	——	0.69
Ferrous oxide	=	2.05	3.67
Magnesia	=	41.03	34.00
Water	=	12.47	12.32
		99.34	99.83

104. *Deweylite* (*Emmons*).

Amorphous, resembling some varieties of gum arabic, or a yellowish or brownish resin. Often incrusting and gradually changing into a feldspar or other minerals, from the decomposition of which in some instances it has been derived. H. = 2–3.5 ; sp. gr. = 2–2.3. Colors whitish, yellowish, brownish, reddish ; translucent. Lustre resinous to waxy ; brittle.

Principally a hydrous silicate of magnesia with the oxygen ratio of $MgO : SiO_2 : H_2O = 2 : 3 : 3$, agreeing with the formula : $4MgO,3SiO_2+6H_2O$.

It has been found massive and in brownish, resinous, botryoidal concretions at Wood's Mine, Lancaster county ; brown massive, resulting from the alteration of feldspar, at the magnesia quarries, West Nottingham township, and near R. Taylor's mill, West Goshen township, Chester county ; also in Radnor township, Delaware county.

The variety from Wood's Mine has been analyzed by G. J. Brush (Dana's Mineralogy, 1854, 236), who found :

Silicic acid	=	43.15
Alumina	=	trace
Magnesia	=	35.95
Water	=	20.25
		99.35

105. *Genthite* (*Dana*).

Amorphous, in delicately hemispherical, mammillary or stalactitic incrustations. H. = 3–4 ; sp. gr. = 2.409. Brittle ; apple-green, yellowish-green. Lustre, resinous ; opaque to translucent. The chemical composition analogous to deweylite.

Upon chromite, associated with and graduating into deweylite, at Wood's Mine, Lancaster county.

I described this mineral, considering it merely as a variety of gymnite (deweylite), as "nickelgymnite," stating the quantity of nickeloxide to be variable and more or less replaced by magnesia, some specimens containing 24.78, others as high as 36.41 per cent. of nickeloxide. (Keller-Tiedemann, Nordam. Monatsbericht, III, 488.)

My analysis of an apple-green variety gave:

Silicic acid	=	35.36
Niccolous oxide	=	30.64
Ferrous oxide	=	0.24
Magnesia	=	14.60
Lime	=	0.26
Water	—	19.09
		100.19

It has also been observed in small quantity at the Red Pit and Low's Mine, Lancaster county.

106. *Kaolinite* (*S. W. Johnson*).

Orthorhombic. If crystallized, in minute rhombic, rhomboidal or hexagonal scales; usually in compact friable or mealy masses. H. = 1–2.5; sp. gr. = 2.4–2.63. Lustre of plates pearly; the massive, from pearly to earthy. White, yellowish, grayish, brownish, reddish-white, &c.

The chemical composition of kaolinite is a hydrous silicate of alumina, with the oxygen ratio of $Al_2O_3 : SiO_2 : H_2O = 3 : 4 : 2$ corresponding with $Al_2O_3, 2SiO_2 + 2H_2O$.

It is found at Tamaqua in scaly crystals, of a white or yellowish color and pearly lustre, and in leathery or rag-like coatings on quartz and of white or yellowish-white color and a silky-pearly lustre, at the coal mines on Mahanoy R. R., in Litchfield township, Schuylkill county. I have analyzed and described (Am. J. of Sc., XXVIII, 251) the scaly mineral as *pholerite;* a similar mineral was subsequently found at Summit Hill, Carbon county, in a cavity of a coal seam, in the form of a scaly brown powder, which after the removal of the coloring iron-compound by chlorhydric acid, became nearly white, of a pearly lustre and soapy feel. A description and analysis is given by Johnson and Blake (Am. J. of Sc., [2] XLIII, 351) in their investigation on kaolinite and pholerite.

The ordinary pulverulent kaolinite or kaolin, of a white or yellowish-white color, is abundantly met with in many parts of southeastern Pennsylvania as the product of the decomposition of orthoclase, and is found almost invariably where the latter occurs; it has been mined to a considerable extent in Concord and Birmingham townships and elsewhere in Delaware county, in New Garden, Kennett and East Nottingham townships, Chester county.

A deposit of white, yellowish, grayish-white, and gray kaolin, which is very free from grit, and which is represented as occurring in abundance, has lately been opened twelve miles north of Douglassville, in Berks county.

The kaolin, as it is found, is rarely pure enough for the manufacture of porcelain, fire-bricks, paper and paper-hangings, and must be washed, many varieties yielding not over twenty to twenty-five per cent. of purified kaolin.

I endeavored to obtain some reliable information as to the annual production of this material, but was not successful.

I have analyzed two samples of kaolin, one from Chestnut Hill, Lancaster county, the other from East Nottingham township, Chester county.

The analyses of kaolinite and kaolin from Pennsylvania gave the following results:

		Kaolinite.		Kaolin.	
		Tamaqua. Mean of two analyses of scales, purified by chlorhydric acid.	Summit Hill. Purified by chlorhydric acid.	Chesnut Hill, Lancaster Co.	E. Nottingham, Chester Co.
		GENTH.	JOHNSON.	GENTH.	GENTH.
Sp. gr.	=	——	2.59	——	——
Silicic acid	=	46.90	45.93	67.1	46.34
Alumina	=	39.60	39.81	20.1	36.32
Ferric oxide	=	——	——	3.9	0.64
Magnesia	=	——	——	0.7	trace
Lime	=	——	——	0.1	0.04
Soda	=	0.17	——	trace	trace
Potash	=	——	——	2.2	0.77
Water	=	13.80	14.02	5.9	13.75
Orthoclase	=	——	——	——	1.04
Quartz	=	——	——	——	1.10
		100.47	99.76	100.0	100.00

The kaolin from Chestnut Hill evidently contains a large admixture of quartz and orthoclase.

All the numerous varieties of clay for pottery, ordinary bricks, fire-bricks, &c., which are found in many localities in the State, the fire-brick clay, especially, in strata in the coal measures, owe their plasticity and other valuable properties, to a great extent, to the kaolinite, which makes up the bulk of the same, more or less contaminated with quartz, oxides of iron, lime, &c.

It is not within the province of this report to speak of them more fully.

I have to mention, however, several peculiar clays, which have been found in this State.

The first occurs at the Ueberoth Zinc Mine near Friedensville, Lehigh county, where it has been discovered by Prof. W. Th. Roepper, who named it "Sauconite," and kindly furnished me with the following description and analyses. It has also been analyzed by John M. Blake (Dana's Mineralogy 1868, 409).

Apparently amorphous, fracture conchoidal; streak brown and shining; colors buff, ochre-yellow, brown, dark gray and white; translucent on thin edges; translucency increased by wetting. When thrown into water emits a crackling sound. H = 1.5; sp. gr. = 2.66–2.70.

The following varieties have been analyzed by Roepper: *a.* pale yellowish-white; *b.* ochre-yellow (after having been dried during one hour at 105° C.); *c.* Blake analyzed a pale-yellow variety:

		a.	*b.*	*c.*
Silicic acid	=	48.94	46.45	41.36
Alumina	=	10.66	7.41	8.04
Ferric oxide	=	3.85	14.28	9.55
Zinc oxide	=	26.95	22.86	32.24
Magnesia	=	——	} 0.97	1.02
Lime	=	2.42		——
Potash	=	——	——	trace
Water	=	7.06	6.73	7.76
		99.88	98.69	99.97

Like all similar minerals, the composition is somewhat variable, owing in part to accidental admixtures, and a replacement of one isomorphous substance by another.

Allowing in the first analysis for a mechanical admixture of 3.45 per cent. of silicic acid (quartz), the oxygen ratio of zinc-oxide and lime (RO), to alumina and ferric oxide (R_2O_3), to silicic acid and water are =1 : 1 : 4 : 1, corresponding with the formula: $3 (RO,SiO_2)+R_2O_3,3SiO_2+3H_2O$.

I believe the sauconite to have as good a claim to be a distinct mineral species, as nine-tenths of the amorphous species, acknowledged as good.

Another peculiar variety of clay, which occurs in considerable quantities at Jones' Mine, Berks county, contains from 2.5 to 10 per cent. of copper.

Several years ago it was mined and exported to England. A small quantity has also been reduced at the Schuylkill Copper Works and in Baltimore.

The clay, which accompanies the limonite deposit at Gen. Trimble's mine, near the White Horse Station, East Whiteland township, Chester county, contains several per cent. of phosphoric acid, probably in the form of wavellite, which at the same locality is found in beautiful specimens.

107. *Halloysite (Berthier).*

Massive, clay-like, earthy. H = 1–2; sp. gr. = 1.8–2.4. Colors, when pure: white, usually yellowish, brownish, grayish, reddish. Lustre dull to waxy, sometimes inclining to pearly. A silicate of alumina with the oxygen ratio for
$Al_2O_3 : SiO_2 : H_2O = 3 : 4 : 3 = Al_2O_3,2SiO_2+3H_2O$.

A yellowish-white variety, of the appearance of cerolite, and frequently called so, has been found, associated with zoisite and euphyllite at the corundum locality on Johnson Patterson's farm near Unionville, Newlin township, Chester county, and was analyzed by Smith and Brush (Am. Journ. Sc. [2] XV, 211). They found:

Sp. gr.	=	2.22
Silicic acid	=	44.50
Alumina	=	25.00
Magnesia	=	7.75
Manganous oxide, Soda and Potash	=	traces
Water	=	22.39
		99.64

A mineral resembling halloysite is found in compact porcelain-like masses at the zinc mines at Friedensville; another halloysite-like mineral occurs with the cobaltiferous wad in Williams township, Northampton county.

An earthy white, yellowish or greenish-white, or yellowish-brown mineral, very smooth, easily cut and polished, which forms seams in the serpentine of West Chester, and contains a large percentage of silicate of alumina may belong here; also a similar mineral occasionally found with chromite at Mineral Hill, Delaware county.

Aquacreptite (Shepard).

Prof. C. U. Shepard (Am. Journ. Sc. [2] XLVI, 256) has described under this name a *new* species, which is found in narrow seams in serpentine at Strode's Mill near West Chester, East Bradford township, Chester county. Massive. H = 2.5; sp. gr. = 2.05–2.08. Yellowish-brown; streak orange-yellow; dull; brittle; fracture small conchoidal. Adheres feebly to the tongue; when thrown into water falls asunder with a crackling noise.

It has been analyzed by Prof. J. H. Eaton, by Mr. Henry C. Humphrey, and by Prof. C. U. Shepard, with results as follows:

		Eaton.	Humphrey	Shepard.
Silicic acid	=	43.03	41.56	41.00
Alumina	=	5.56	6.71	4.00
Ferric oxide	=	12.30	12.45	13.30
Magnesia	=	19.58	not det'd	17.60
Water	=	17.40	16.00	23.00
		97.87		98.90

A similar mineral of a brown color, which crackles and falls to powder under water, occurs in the limestone quarries at Marble Hall, Montgomery county.

108. *Damourite (Delesse), Corundellite (Silliman, in part).*

Probably orthorhombic with monoclinic habit. Sometimes in groups of hexagonal plates, globularly arranged or in smaller hexagonal crystals; usually in aggregates of fine scales, often crypto-crystalline and with a fibrous and radiating structure;

slaty. H. = 2–3.5 ; sp. gr. = 2.76–2.86. Color silvery white, grayish-white, sometimes very delicate greenish-white and green; rarely yellowish oil-green. Lustre of scaly varieties pearly; of the fibrous, pearly-silky; also dull. Damourite is essentially a hydrous muscovite; a hydrous silicate of alumina and potash, with the oxygen ratio of $K_2O : Al_2O_3 : SiO_2 : H_2O = 1 : 9 : 12 : 2$.

A great variety of forms of damourite—*all resulting from the alteration of corundum*, and many in their aggregate of fine scales or fibres still retaining the form of the original mineral, are found at Unionville, Newlin township, Chester county.

Here the cockscombed-shaped crystalline groups of one inch broad and three-quarters of an inch thick and crystalline plates of three to four inches in width, have been observed, together with all the other varieties.

They have been investigated by S. P. Sharples (Am. Jour. Sc. [2] XLVII, 319) and more fully by myself in my paper on the alteration of corundum (*l. c.*), where my own, Dr. G. A. Koenig's and Mr. Thomas M. Chatard's analyses of the different varieties are published. An analysis of a granular mica, from Unionville, by J. D. Darrach (Dana's Mineralogy, 1850, 357), belongs here, the percentage of potash evidently being too low:

		1. Granular.	2. White & greenish-white crystals, mean.		3 Silver-gray scales, pseudom.	4 White scaly radiating, pseudom.
		DARRACH.	SHARPLES,	KOENIG.	GENTH.	GENTH
Sp. gr.	=	——	2.87	2.851	2.843	——
Silicic acid	=	46.75	43.56	43.03	45.57	45.86
Alumina	=	39.20	38.16	39.06	34.83	37.65
Ferric oxide	=	trace	——	1.48	2.94	0.59
Magnesia	=	1.02	——	0.30	0.83	0.55
Lime	=	0.39	——	trace	0.40	0.31
Lithia	=	——	——	trace	trace	trace
Soda	=	——	——	0.58	0.87	0.80
Potash	=	6.56	10.81	10.05	10.16	10.40
Water	=	4.90	5.64	5.40	5.30	4.74
		98.82	98.17	99.90	100.90	100.90

9

		5. Greenish-white, massive. pearly.	6. White, pseudo-fibrous.	7. Yellowish oil-green crypto-crystalline, pseudomorphous.	8. Yellowish oil-green crypto-crystalline, waxy.
		KOENIG.	KOENIG.	GENTH.	CHATARD.
Sp. gr.	=	2.857	2.832	2.779	2.760
Silicic acid	=	45.73	45.73	46.98	46.60
Alumina	=	37.10	36.30	35.13	32.39
Ferric oxide	=	1.30	0.83	0.61	2.54
Magnesia	=	0.34	0.54	1.32	2.01
Lime	=	trace	0.74	0.13	trace
Lithia	=	trace	trace	trace	trace
Soda	=	0.88	0.58	0.76	0.54
Potash	=	10.50	10.49	10.74	10.39
Water	=	4.48	5.17	4.77	4.81
		100.33	100.38	100.44	99.28

Damourite has also been found with nuclei of unaltered corundum in Aston township, Delaware county.

The white micaceous mineral, which accompanies the cyanite of White Horse, below Darby, is probably damourite.

The micaceous mineral, which occurs in implanted crystals, in minute tuft-like aggregations, the crystals rarely over a line in diameter, upon dolomite in the Poorhouse quarry, Chester county, belongs undoubtedly to damourite.

It has been analyzed by Smith and Brush (Am. Jour. Sc. [2] XVI, 47), who found :

Silicic acid	=	45.50
Alumina	=	34.55
Ferric oxide	=	trace
Magnesia	=	1.08
Lime	=	2.31
Soda	=	2.35
Potash	=	8.10
Water	=	5.40
		99.29

A large portion of the magnesia and lime is (as S. and B. state) doubtless due to dolomite.

I have to mention in connection with this species several varieties of the so-called *lesleyite*, from Unionville, which I consider (*l. c.*) alterations of corundum into fibrolite and a subsequent change of this into damourite, a portion of the original corundum still remaining unconverted.

A white variety has been analyzed by S. P. Sharples (Am. Jour. Sc. [2] XLVII, 319), Mr. Chatard and myself; the reddish variety by C. W. Roepper (quoted in Sharples' paper), Dr. Koenig and myself.

Omitting Roepper's analysis, in which the free corundum had not been separated from the silicic acid, the following results were obtained:

		White.			Reddish.	
		GENTH.	CHATARD.	SHARPLES.	GENTH.	KOENIG.
Sp. gr.	=	——	3.200	3.203	——	3.059
Silicic acid	=	32.32	32.32	33.59	31.96	31.90
Alumina	=	56.43	55.23	55.41	} 56.85	54.09
Ferric oxide	=	0.29	——	——		0.51
Magnesia	=	0.38	} 0.73	——	0.13	0.34
Lime	=	0.32		——	0.13	——
Lithia	=	faint traces		——	faint traces	
Soda	=	0.32	0.64	——	0.35	1.01
Potash	=	7.31	7.82	7.43	7.83	8.71
Water	=	4 01	3.86	4.30	4.09	4.20
		101.38	100.60	100.73	101.34	100.76

At several localities occur micaceous minerals, pseudomorphous after others, for instance at Dutton's mill, Delaware county, after cyanite; in Marple township, Delaware county, after andalusite; and near Texas, Lancaster county, after tourmaline.

It is probable that these pseudomorphous micaceous minerals are *damourite.*

Recent investigations have repeatedly pointed out the importance of damourite as an essential constituent of many rock masses.

But very little has been done yet with the Pennsylvania rocks, the few analyses, which I have made or had made in the Laboratory of the University of Pennsylvania, prove not only the roofing slates of Lancaster county, which cross the Susquehanna River at Peachbottom, to contain a considerable percentage of damourite, but also many of the slates, occurring in the lower Silurian limestones.

The following are the results of the analyses:

a. Roofing slate from opposite Peachbottom, in Lancaster county, by myself; *b.* A grayish, reddish-white slate, with a delicate silky lustre, from Krömmlich and Lichtenwallner's Mine, in Fogelsville, Lehigh county, by myself; *c.* A grayish-white slate, with silky lustre, from Thomas & Co.'s Mine, at Hensingerville, near Alburtis, Lehigh county, by Mr. Sydney Castle; *d.* A dull aluminous slate of a brownish pale-gray color, from near East Penn junction, by Mr. Pedro G. Salom, and *e.* A similar gray slate, from the same neighborhood, also by Mr. Pedro G. Salom.

		a	*b*	*c*	*d*	*e*
Silicic acid	=	60.32	49.92	45.40	59.30	39.80
Carbonic acid	=	——	——	——	——	14.40
Ferric oxide	=	——	0.91	5.06	} 30.30	2.40
Alumina	=	23.10	34.06	24.69		23.95
Ferrous oxide	=	7.05	——	——	——	——
Magnesia	=	0.87	1.77	13.56	trace	1.94
Lime	=	——	0.11	trace	trace	9.85
Lithia	=	——	trace	——	——	——
Soda	=	0.49	0.74	0.27	1.51	0.52
Potash	=	3.83	6.94	5.85	6.24	3.34
Water	=	4.08	6.52	4.80	4.70	6.00
Graphite	=	0.56	——	——	——	——
Pyrite	=	0.09	——	——	——	——
		100.39	100.97	99.63	102.05	102.20

Taking for comparison the typical damourite, which contains 11.77 per cent. of potash, the above slates have an admixture of the following quantities of damourite:

a	=	35.08	per cent.
b	=	55.40	"
c	=	49.70	"
d	=	53.02	"
e	=	28.30	"

The last slate contains a large quantity of carbonates of lime and magnesia.

109. *Euphyllite* (*Silliman, Jr.*).

This species, which occurs associated with corundum, zoisite and tourmaline, at Unionville, Newlin township, Chester county, was distinguished in 1849 by Prof. Silliman, who gave a description and preliminary analysis by Crooke (Am. Journ. Sc. [2] VIII, 381).

It has the structure of mica, but the laminæ are less easily separable. H = 3.5–4.5 ; sp. gr. = 2.83 (S. & B.) –2.963–3.008 (Sill.). Colorless to white, the sides faint grayish sea-green or whitish, transparent to translucent; also opaque; lustre of cleavage planes bright pearly, inclining to adamantine.

It has been analyzed by Crooke (*l. c.*), subsequently by Erni (Dana's Min., 1850, 362), and Garrett (Dana's Min., 1850, 362), and finally by Smith and Brush (Am. Journ. Sc. [2] XV, 209).

The analyses of Crooke, Erni and Garrett having been proved by these more recent investigations to be erroneous, I give only the latter.

Smith and Brush found:

Silicic acid	=	40.29	39.64	40.21	40.96
Alumina	=	43.00	42.40	41.50	41.40
Ferrous oxide	=	1.30	1.60	1.50	1.30
Magnesia	=	0.62	0.70	0.78	0.70
Lime	=	1.01	1.00	1.88	1.11
Soda	=	5.16	5.16	4.26	4.26
Potash	=	3.94	3.94	3.25	3.25
Water	=	5.00	5.08	5.91	6.23
		100.32	99.52	99.29	99.21

A micaceous mineral, resembling euphyllite, is found with crystals of gray corundum in the granular albite near Unionville, Newlin township, Chester county.

Owing to want of material only two *partial* analyses were made by Dr. Geo. A. Koenig; it is probably the same of which a *partial* analysis, by S. P. Sharples, is published in Dana's Mineralogy, and identical with a similar micaceous mineral, associated with the emery of Asia Minor, analyzed by J. L. Smith (Am. Journ. Sc. [2] XI, 62):

		Koenig.		Sharples
		1.	2.	3.
Silicic acid	=	41.45	41.46	43
Alumina	=	42.30	42.19	40
Soda Potash	= =	} [7.32]	[7.45]	7–8
Water	=	6.66	8.90	——
Corundum	=	2.27	——	——
		100.00	100.00	

110. *Hisingerite* (*Berzelius*).

This mineral has been found at the Gap Mine, Lancaster county, by Mr. Theo. D. Rand, who gave a description and analysis of it (Proc. Ac. Nat. Sc., Phil., 1872, 304).

It occurs in black, amorphous masses, of a lustre between resinous and vitreous; streak brown; fracture conchoidal; brittle. H = 2.5-3; sp. gr. = 2.11.

The analysis, omitting 1.13 per cent. of gangue, gave:

Water expelled at 100° C	=	14.30	=	24.19
" " redness	=	9.89		
Silicic acid			=	35.40
Ferric oxide			=	27.46
Ferrous oxide			=	12.53
				99.58

---

3. *Subsilicates*.

Under subsilicates are classed peculiar groups of minerals, crystallizing in the rhombohedral, rhombic and monoclinic systems, all having a very eminent basal cleavage, which gives them a foliated structure. They are mostly of a dark blackish-green or brown color. Although we have of many very valuable analyses, their chemical constitution is not yet settled, because the amount of water, which they contain, appears to be present in *two* conditions: in the first place the hydrogen of the water, replacing basic metals, and secondly, water of crystallization.

Prof. Josiah P. Cooke, Jr., has lately published a very important paper on vermiculites, &c. (Proc. Am. Ac., Dec. 9, 1873), in addition to previous investigations on the crystallographic examination of some American chlorites (Am. Journ. Sc. [2] XLIV, 201), which together throw a great deal of light on the constitution of these minerals.

Many of the species can only be determined by analysis, combined with optical investigations.

The following are the species recognized in Pennsylvania, with the localities where they have been found:

111. *Jefferisite* (*Brush*).

Probably orthorhombic. In broad crystals and crystalline plates, sometimes several inches across, with eminent basal cleavage, the cleavage planes marked by the crossing of lines at angles of 60° and 120°. H = 1.5; sp. gr. = 3.30. Color, dark yellowish-brown and brownish-yellow. When heated to 300° C. it exfoliates very remarkably. The chemical constitution is according to Cooke (*l. c.*) that of a hydrous biotite.

It occurs in veins in serpentine three miles south of West Chester, in Westtown township, Chester county, also in hexagonal crystals of about one to two inches broad, near Unionville, Newlin township, Chester county.

A mineral in small brownish-yellow crystalline plates and exfoliating like jefferisite, occurs at the soapstone quarries near Lafayette, Montgomery county; similar minerals have been found in Middletown township, Delaware county, and in small plates at Texas, Lancaster county.

The West Chester jefferisite has been analyzed by Prof. Brush (Am. Journ. Sc. [2] XXXI, 370), Dr. Koenig and Mr. Chatard (my paper on Corundum, *l. c.*), with the following results:

		Brush.	Koenig.	Chatard.
Silicic acid	=	37.10	33.35	34.40
Alumina	=	17.57	17.78	16.63
Ferric oxide	=	10.54	7.32	8.00
Ferrous oxide	=	1.26	2.11	2.11*
Magnesia	=	19.65	19.26	19.30
Lime	=	0.56	——	——
Soda	=	trace	not determined	
Potash	=	0.43	"	"
Water	=	13.76	19.87	19.03
		100.87	99.69	99.47

None of the other varieties has been analyzed.

112. *Hallite* (*Cooke*).

First described by A. R. Leeds (Frankl. Inst. Journ. [3] LXII, 70). In large rough six-sided prisms with easy micaceous cleavage. Color green and brownish-yellow. Ex-

* Taken from Dr. Koenig's determination.

foliates like jefferisite. The chemical constitution is according to Prof. Cooke that of a hydrous phlogopite. It is found in East Nottingham township, Chester county, three miles south of Oxford, in nests or pockets of a greenish or yellowish steatitic earth in the serpentine.

It has been analyzed by C. E. Munroe (Cooke's paper, *l. c.*).

		Green variety. Mean of 2 analyses.	Yellow variety. Mean of 2 analyses.
Silicic acid	=	35.89	35.26
Alumina	=	7.45	7.58
Ferric oxide	=	8.78	9.68
Ferrous oxide	=	1.13	0.32
Magnesia	=	31.45	31.51
Potash	=	0.46	0.61
Water	=	14.33	14.78
		99.49	99.74

A very similar mineral in green and yellowish-brown foliated masses, which is probably the same species, has been found near Lenni, Delaware county.

113. *Penninite* (*Dana*).

Rhombohedral with highly perfect basal cleavage; crystals often tabular and in crested groups. Massive in fine scales, often in fibrous aggregations, compact crypto-crystalline.

H. = 2–2.5; sp. gr. = 2.6–2.85. Color, various shades of green, grass-green, grayish and deep olive-green; also deep violet, pink, grayish-red, occasionally yellowish and silver-white. Dichroic.

A hydrous silicate of alumina, partly replaced by chromic and ferric oxides, and of magnesia, partly replaced by ferrous oxide. The oxygen ratio of the bases and silicic acid varying from 4 : 3 to 5 : 4.

The penninite variety "kaemmererite" is found at Wood's Mine in beautiful green and violet crystals, more frequently in pink or peach-blossom colored scaly-granular or scaly-fibrous aggregations; sp. gr. = 2.617–2.62.

It has been analyzed by T. H. Garrett (Am. Jour. Sc. [2] XV, 332); a pink scaly variety, by myself (Proc. Ac. Nat. Sc., Phil., 1852, 121); Hermann (Jour. Prakt. Chem. LIII, 1); Smith and Brush (Am. Jour. Sc. [2] XVI, 47); and John B. Pearse (Am. Jour. Sc. [2] XXXVII, 222).

Omitting Garrett's analysis, which has not been confirmed by later investigations. the results are as follows:

		Genth. Mean of 2 anal.	Hermann.	Smith and Brush		Pearse.	Pearse.
Silicic acid	=	33.20	31.82	33.26	33.30	31.86	31.31
Alumina	=	11.11	15.10	10.69	10.50	13.75	12.84
Chromic oxide	=	6.85	0.90	4.78	4.67	2.15	2.98
Ferric oxide	=	1.43	4.06	1.96	1.60	——	——
Ferrous oxide	=	——	——	——	——	2.31	2.46
Niccolous oxide	=	——	0.25	——	——	0.22	0.45
Magnesia	=	35.54	35.24	35.93	36.08	34.90	35.02
Lime	=	——	——	——	——	1.27	0.82
Lithia	=	} 0.28	——	——	——	——	——
Soda	=		——	} 0.35	0.35	} ——	——
Potash	=	0.10	——			——	——
Water	=	12.95	12.75	12.64	13.25	13.98	13.20
		101.46	100.12	99.61	99.75	100.44	99.08

Scaly-granular and scaly-fibrous pink or peach-blossom colored kaemmererite is also found at Moro Phillips' Chrome Mine, in Marple township, Delaware county, and at Scott's Chrome Mine, West Nottingham township, Chester county.

114. *Ripidolite* (*v. Kobell*).

Monoclinic forms, with an eminently basal cleavage, crystals often tabular, also oblong and rhombohedral in aspect; frequently grouped in rosettes; also massive, coarse scaly-granular to fine granular and earthy.

H. = 2–2.5; sp. gr. = 2.65–2.77. Color, deep grass-green to olive-green, also sometimes rose-red. Strongly dichroic.

Ripidolite resembles penninite in its chemical composition; the oxygen ratio of $RO : R_2O_3 : SiO_2 : H_2O = 5 : 3 : 6 : 4$.

It is intimately associated, and forms frequently compound crystals of great beauty with kaemmererite; sometimes crystals of the latter being imbedded in ripidolite and the reverse; upon chromite, at Wood's Mine, Lancaster county. Plates of green ripidolite are sometimes found implanted in dolomite, at the same locality.

Magnificent crystals of very large size have been discovered at

Ingram's Quarry and Brinton's Quarry, three miles south of West Chester, by Mr. W. W. Jefferis, whose cabinet contains many very fine specimens of this and many other Pennsylvania species. The West Chester ripidolite has been called clinochlore, by W. P. Blake, who described it in 1851 (Am. Jour. Sc. [2] XII, 339).

It has been analyzed by W. J. Craw (Am. Jour. Sc. [2] XIII, 222); and lately by Ed. F. Neminarz, especially with reference to the state of the oxydation of the iron in this mineral (Tschermak's Mineralogische Mittheilungen, 1874, 177).

It contains:

		CRAW.	CRAW.	NEMINARZ.
Silicic acid	=	31.34	31.78	31.08
Alumina	=	17.47	}	18.85
Chromic oxide	=	1.69	} 22.71	1.09
Ferric oxide	=	3.85	}	1.55
Ferrous oxide	=	——	——	2.33
Magnesia	=	33.44	33.64	33.50
Lime	=	——	——	0.81
Water	=	12.60	12.60	11.53
		100.39	100.73	100.74

It also occurs at Patterson's Farm, near Unionville, Newlin township, Chester county, and probably at Jones' Mine, Berks county.

115. *Prochlorite* (*Dana*).

Its crystalline form is probably hexagonal, with eminent basal cleavage; often in divergent groups, fan-shaped or spheroidal; from large folia to coarse or fine scaly-granular and sometimes scaly-fibrous; massive, frequently crypto-crystalline and earthy. H. = 1–2; sp. gr. = 2.78–2.96. Color grass-green, olive-green, blackish-green.

It contains the same constituents (excepting chromic oxide) as the two preceding species; the oxygen ratio of $RO : R_2O_3 : SiO_2 : H_2O = 7 : 6 : 8 : 6$, corresponding with: $4RO,SiO_2 + 3RO,2R_2O_3 + 6H_2O$.

It is found in considerable quantities as the result of alteration of corundum, frequently containing yet a nucleus of the latter; at Unionville, in foliated dark-green masses, also in fine scaly and slaty masses, and light olive-green, exceedingly finely granular; almost impalpable. The last named variety usually occurs as an indirect alteration of corundum, this mineral first being changed into pseudo-fibrous damourite, and the latter into prochlorite; minute crystals of tourmaline in the damourite are, on entering the chloritic mineral, also converted into prochlorite, retaining their original form.

A. R. Leeds (Am. Journ. Sc. [3] VI, 25) has analyzed it and I give his analysis, together with two others, made by Th. M. Chatard (my paper on Corundum, *l. c.*); Dr. Koenig has analyzed two varieties of decomposed chlorite from specimens, which still contain a nucleus of unaltered corundum (*l. c.*).

A variety of an olive-green granular, scaly prochlorite, which occurs in seams of several inches in thickness in the slaty serpentine of Rose's quarry, opposite Lafayette in Montgomery county, has been analyzed in the Laboratory of the University of Pennsylvania, by Mr. Sydney Castle, whose analysis I shall give below.

Fine scaly dark-green chlorite (? prochlorite) is found in West Goshen, Chester county.

In many parts in the State chlorite and chloritic minerals occur in considerable quantities but of none is it ascertained whether they belong to prochlorite or another species.

Crystallized chlorite has been found in dark bottle-green foliated and mammillary aggregations in hornblendic rocks near the Falls of the Schuylkill; it occurs foliated massive, and abundantly as chlorite schist in the soapstone quarries near Lafayette, Montgomery county, and north and south of it as far as these rocks extend. In the neighborhood of Texas, Lancaster county, massive, scaly-granular chlorite with magnetite) occurs abundantly.

It is found compact and crystalline at Easton.

Frequently it is associated with magnetic iron ores, in the South Mountains, for instance, at Seissholzville, Topton, Fritz's Island and other mines in the neighborhood of Reading.

With the allanite near Bethlehem a variety of a very dark blackish-green color and crypto-crystalline, scaly structure is found, resembling thuringite, which, in the absence of investigation, I will merely mention here.

The analyses of the finely granular, light olive-green prochlorite from Unionville, gave as follows:

		Leeds.	Chatard.	
Silicic acid	=	30.62	29.43	29.59
Alumina	=	21.73	22.08	22.18
Ferric oxide	=	0.42	1.41	1.33
Ferrous oxide	=	5.01	5.64	5.77
Magnesia	=	29.69	28.46	28.54
Lithia	=	0.11	not determined	
Soda	=	0.14	"	"
Water	=	12.26	12.40	12.40
		99.98	99.42	99.81

The analyses of the decomposed prochlorites from the same locality, by Dr. Koenig, gave for the :

		Brownish-red.	Yellowish-gray.
Silicic acid	=	31.35	32.80
Alumina	=	21.58	26.07
Ferric oxide	=	14.17	9.80
Magnesia	=	16.67	17.70
Water	=	14.45	13.75
		98 22	100.12

The analysis of a scaly-granular, dark-green chloritic mineral from Rose's serpentine quarry, opposite Lafayette, Montgomery county, has been made at my request in the Laboratory of the University of Pennsylvania, by Mr. Sydney Castle, who found :

Silicic acid	=	38.80
Alumina	=	4.54
Ferric oxide	=	3.97
Ferrous oxide	=	3.63
Magnesia	=	35.78
Water	=	13.40
		100.12

This composition shows a close resemblance to that of serpentine; the oxygen ratio of

$RO : R_2O_3 : SiO_2 : H_2O = 1 : 0.22 : 1.36 : 0.79$. It requires fuller investigation.

116. *Pattersonite* (*Lea*).

This name was given by Isaac Lea (Proc. Ac. Nat. Sc. Phila. April 9th, 1867) to a chloritic mineral, which occurs with lesleyite at Unionville.

It usually presents triangular plates, sometimes joined to make masses like a tetrahedron; cleavage eminently basal; laminæ not flexible, slightly translucent; color blueish-gray; lustre pearly, inclining to submetallic.

S. P. Sharples (Am. Journ. Sc., [2] XLVII, 320) published an analysis of it, but as Brush proved it to be erroneous (Suppl. to Dana's Mineralogy, 1872, 18), I have made two new ones, but with material, slightly contaminated by oxydation (Corundum, *l. c.*), and found as follows:

		Nearly pure.	Purest.
Spec. grav.	=	——	2.810
Silicic acid	=	29.89	29.90
Alumina	=	} 30.87	27.59
Ferric oxide	=		3.12
Ferrous oxide	=	9.17	9.17
Magnesia	=	17.53	17.10
Lithia	=	trace	trace
Soda	=	0.83	0.58
Potash	=	2.41	2.33
Water	=	11.60	11.51
		102.30	101.30

This mineral is allied to thuringite.

117. *Chloritoid* (*G. Rose*).

This mineral, which in Siberia, Asia Minor and elsewhere, is so frequently associated with corundum, is mentioned in Dana's Mineralogy as occurring at Unionville, Pa. I have not been able to recognize it amongst the chloritic minerals, which occur associated with corundum at this locality.

The variety "phyllite" appears to occur in small greenish-black scales in the siliceous lower Silurian limestone at the Pequea Mine, Lancaster county; also at Conshohocken, Montgomery county.

118. *Margarite* (*Fuchs*).

This very interesting mineral, generally resulting, as far as our present information goes, from the alteration of corundum, occurs in crystals, belonging to the orthorhombic system, also in crystalline, foliated masses of eminent basal cleavage, coarse and fine scaly, pseudo-fibrous, crypto-crystalline. H = 3.5–4.5 ; sp. gr. = 2.99–3.08. Color white, frequently with a delicate pink hue; also grayish, greenish and brownish-white. Lustre pearly on cleavage planes. The chemical composition is that of a hydrous silicate of alumina and lime, the latter more or less replaced by soda and potash. The oxygen ratio of

$RO : Al_2O_3 : SiO_2 : H_2O = 1 : 6 : 4 : 1.$

Corundum crystals from Unionville have frequently a coating of the crypto-crystalline variety, or are even completely altered into the same, occasionally leaving a nucleus of the original mineral. I have analyzed this variety (Corundum, *l. c.*); see below (1).

A similar variety with nuclei of corundum has been found on the north slope of Mineral Hill, Delaware county.

A massive compact or finely granular, white, somewhat ferruginous, crypto-crystalline variety, accompanying granular corundum at Unionville, has been analyzed by me (2).

A laminated white, very faintly pinkish variety, occurs rarely with granular corundum and tourmaline at Unionville; it has also been analyzed by me (3).

A scaly and somewhat laminated, granular aggregate of white margarite, with a faint reddish hue, associated with black tourmaline, from Unionville, has been analyzed by Dr. Koenig (4).

A very peculiar, compact variety of a grayish-white or brownish-white color, waxy lustre and crypto-crystalline structure, also from Unionville, has been analyzed by Thos. M. Chatard (5). It easily breaks into fragments and frequently contains particles of unaltered corundum.

W. J. Craw (6) (Dana's Mineralogy, 1850, 362) and C. Hartshorne (*Ibid*) (7), have also analyzed margarite from Unionville.

Very interesting, usually fine scaly, pearly margarite, often with nuclei of bronze-brown corundum, occurs at Village Green, Aston township, Delaware county. Several samples have been analyzed by W. J. Craw (Am. Journ. Sc., [2] VIII, 379).

UNIONVILLE MARGARITES.

		1. GENTH	2. GENTH.	3. GENTH.	4. KŒNIG.	5. CHATARD.	6. CRAW.	7. HARTSHORNE.
Spec. grav.	=	3.012	3.047	——	——	3.00	——	——
Silicic acid	=	32.19	30.45	30.70	31.48	31.29	29.99	32.15
Alumina	=	49.62	50.86	49.33	49.01	47.24	50.57	54 28
Ferrous oxide	=	0.91	0.42	0.39	0.52	0.85	——	trace.
Magnesia	=	0.41	0.37	0.76	0.54	0.88	0.62	0.05
Lime	=	7.81	12.13	11.86	10.70	10.86	11.31	11.36
Lithia	=	trace	trace	0.36		trace	——	
Soda	=	4.78	1.72	0.96	1.34	2.66	1.62	Not d't
Potash	=	0.57	0.25	0.65		0.24	0.85	
Water	=	3.93	4.48	5.91	3.94	5.92	5.14	0.50(?)
Corundum	=	——	trace	——	2.00	——	——	——
		100.22	100.68	100.92	99.53	99.94	100.10	

VILLAGE GREEN MARGARITES, BY W. J. CRAW.

Spec. grav.	=	2.995	——	——	——
Silicic acid	=	32.311	31.060	31.261	30.18
Alumina	=	49.243	51.199	51.603	51.40
Magnesia	=	0.298	0.283	0.499	0.72
Lime	=	10.663	9.239	10.146	10.87
Soda	=	2.215	2.969	1.221	2.77
Potash	=				
Water	:	5.270	5 270	5.270	4.52
		100.000	100.020	100.000	100.46

Several undetermined silicates occur in the State, which I have not mentioned on account of the uncertain characters of the same; however, I shall take an opportunity of examining them at a future day.

II. *Tantalates—Columbates.*

119. *Columbite.*

Only one crystal of this *rare* mineral was found many years ago in Nivin's quarry, New Garden township, Chester county, by Mr. Sam'l Tyson, of King of Prussia, Montgomery county, who has it preserved in his cabinet. It is an imperfect crystal, about two inches long, a little over one inch broad and not quite half an inch thick. It has a black color.

A chemical examination proved it to be a columbate of iron and manganese.

III. *Phosphates, Arsenates, Antimonates, Vanadates and Nitrates.*

a. Anhydrous.

120. *Apatite* (*Werner*).

Crystallizes in hexagonal prisms, with basal plane, usually in combination with other prismatic and pyramidal planes; also massive. None of the fibrous, columnar, &c., varieties have been observed in Pennsylvania.

H. = 5; sp. gr. = 2.92–3.25. Colorless, sea-green, blueish-green, greenish-yellow and yellow. Its chemical composition is fluoride of calcium and phosphate of lime:
$CaFl_2 + 3CaO,P_2O_5$; a portion of the fluorine usually substituted by chlorine.

Apatite is found in blueish-green crystals in the gneissic rocks in Southeastern Pennsylvania.

At McKinney's quarry, near Germantown, it occurs sometimes in highly polished hexagonal prisms of a blueish-green color, about three and a quarter to one inch in length; more commonly in large rough, very friable crystals, at times two inches in diameter, mostly implanted in massive orthoclase; it is also found at Megargee's paper-mill, and in the gneissic rocks of Frankford; in small crystals at several places in West Philadelphia, and at Gray's Ferry; in a similar manner it occurs in hexagonal prisms at Leiperville; at Beattie's mill, Springfield township, Delaware county; and Unionville, Chester county.

Very fine yellowish-green, transparent six-sided prisms and masses of a conchoidal fracture, are found in the soil on Wm. Jackson's farm, near Penn's Meeting House, in London Grove township; in yellowish-green hexagonal prisms it is found in granular limestone, in Bernard's quarry, West Marlborough township, and in green hexagonal crystals, at Nivin's quarry, New Garden township, Chester county.

In blueish hexagonal crystals and grains, also massive it has been met with in Van Arsdale's limestone quarries, near Feisterville, Bucks county.

In very small, white hexagonal crystals it occurs with magnetite, at Jones' Mine, Berks county;

And in straw- and honey-yellow hexagonal prisms with pyramid and basal planes with crystallized dolomite, in Prince's soapstone quarries, near Lafayette, Montgomery county.

There exists only one analysis of Pennsylvania apatite, by H. Seybert (Jour. Ac. Nat. Sc., Phil., II, 144), of that from London Grove township, but being made over fifty years ago, before accurate methods of separation of phosphoric acid were known, it is of no other than historic value.

Apatite does not occur in sufficiently large masses in the State to be used as a material for artificial fertilizers.

121. *Pyromorphite* (*Hausmann*).

Crystallizes in hexagonal prisms with basal plane, sometimes slightly modified by pyramids and secondary prisms. Crystals usually small, also imperfect, cavernous or barrel-shaped. Often in globular, reniform and botryoidal aggregations; also fibrous and granular. H = 3.5–4; sp. gr. = 6.5–7.1. Colors of those found in Pennsylvania different shades of green, from light yellowish-green to deep olive-green. Lustre resinous. The chemical composition is chloride of lead and phosphate of lead = $PbCl_2 + 3PbO,P_2O_5$. Frequently the chloride of lead is partly replaced by fluoride of calcium, and part of the phosphoric acid by arsenic acid.

It occurs in beautiful groups of various shades of green, some of a very deep, almost black olive-green color; also pea-green, leek-green and greenish-yellow, near Phœnixville, especially at the Wheatley, Brookdale, Chester county and Charlestown Mines. The crystals are sometimes of considerable size, some of half an inch in diameter have been found; usually hexagonal prisms, but occasionally modified by pyramids; often cavernous. It also occurs in columnar, botryoidal, plumose and granular masses. A very dark green variety, examined by J. L. Smith, in his "Minerals of the Wheatley Mine" (Am. Journ. Sc. [2] XX,

248), had a sp. gr. of 6.94. No analysis has been made, but I found the dark green color to be due to chromic oxide.

Large quantities of pyromorphite, from the Chester county silver-lead Mines, have been smelted.

Small pale crystals and botryoidal aggregations and coatings have also been found at the Ecton Mine, near Shannonville, Montgomery county.

122. *Mimetite* (*Haidinger*).

The crystalline form similar to that of pyromorphite, but the crystals more frequently modified by pyramids. It has rarely been found at the Wheatley and Charlestown Mines, near Phœnixville.

The crystals are sometimes nearly colorless, usually yellow and greenish-yellow. In hair-like hexagonal, sometimes in barrel-shaped crystals of a greenish-yellow, or yellow flat hexagonal tables upon dark green pyromorphite, also in straw-yellow hexagonal prisms. Some of the crystals are in part pyromorphite, these two species changing into one another without a line of demarcation. When pure it is a combination of chloride of lead and arsenate of lead = $PbCl_2 + 3PbO,As_2O_5$.

A lemon-yellow variety analyzed by J. L. Smith (*l. c.*) gave:

Spec. grav.	=	7.32
Arsenic acid	=	23.17
Chlorine	=	2.23
Plumbic oxide	=	67.05
Lead	=	6.99
Phosphoric acid	=	0.14
		99.74

123. *Descloizite* (*Damour*).

This exceedingly rare mineral has been found in very minute quantity, in dark colored crystalline crusts, on quartz and ferruginous clay, at the Wheatley Mine. When magnified it is found to consist of minute lenticular crystals, grouped in small botryoidal masses; their color is dark purple, almost black; in transmitted light, almost hyacinth-red and translucent, streak yellow. J. L. Smith (*l. c.*) has made an analysis of an impure variety, contaminated with wulfenite and other impurities. After their subtraction, he finds:

Vanadic acid	=	33.9
Plumbic oxide	=	66.1
		100.0

The conclusion of Dr. Smith that the Wheatley Mine vanadate of lead is descloizite, is corroborated by a letter to him from Descloizeau (Am. Journ. Sc. [2] XLVIII, 137), where he gives the results of a few measurements, showing the form to be a rhombic table with an angle 64°–65°, corresponding with the prism adopted for descloizite.

b. *Hydrous.*

124. *Vivianite* (*Werner.*)

It crystallizes in the monoclinic system, in prismatic crystals frequently highly modified. Cleavage eminent clinodiagonally; lustre vitreous, upon the cleavage plane strongly pearly. H. = 2; sp. gr. = 2.6–2.7. Rarely almost colorless, inclining to greenish- or blueish-white; mostly various shades of blue to dark steel-blue, resulting from a partial oxydation. In its pure state a hydrous ferrous phosphate = $3FeO,P_2O_5 + 8H_2O$, but mostly containing more or less ferric oxide.

Very fine, almost colorless, slender, prismatic crystals of nearly 1½ inches in length, and small modified steel-blue crystals have rarely been found at the Gap Mine, Lancaster county.

125. *Libethenite* (*Breithaupt*).

An orthorhombic hydrous phosphate of copper, usually of a dark olive-green color.

On the authority of Mr. Chas. M. Wheatley (priv. com.), I give the Perkiomen Mine near Shannonville, as a locality.

It has not been observed elsewhere.

126. *Pseudomalachite* (*Hausmann*).

Another hydrous phosphate of copper, either of hemihedral orthorhombic or of monoclinic forms, and usually of a dark emerald—or verdegris-green color.

It has occurred at the Perkiomen mine, but rarely.

127. *Wavellite* (*Babbington*).

Orthorhombic, usually very minute crystals; mostly in hemispherical, globular, botryoidal and stalactitic masses, with a radiated structure. H. = 3.25–4; sp. gr. = 2.316–2.337. White, grayish, yellowish, greenish-white. Lustre vitreous, inclining to resinous and pearly. It consists of a hydrous phosphate of alumina with the oxygen ratio for Al_2O_3; P_2O_5 : H_2O = 9 : 10 : 12.

The most beautiful varieties of wavellite are found in Gen. Trimble's iron mine, near White Horse Station, on Chester Valley Railroad in East Whiteland township, Chester county.

Here it occurs in minute, colorless, prismatic crystals, aggregations of crystals and radiating tufts of acicular crystals; also in white, yellowish and grayish-white opaque, sometimes botryoidal and stalactitic, radiating masses. The wavellite is often coated with pearly gibbsite. (This locality is often, but erroneously called " Steamboat.")

I have analyzed the stalactitic variety, (Am. Journ. Sc. [2] XXIII, 423); a less pure variety has lately been re-examined by Hermann (Bull. Soc. Impér. Nat. Moscow, No. 4, 496).

		Genth.	Hermann.
Spec. gr.	=	——	2.30
Phosphoric acid	=	34.68	32.70
Alumina	=	36,67	35.83
Water	=	28.29	28.39
Limonite	=	0.22	——
Ferric oxide	=	——	3.08
Fluorine	=	trace	trace
		99.86	100.00

In small hemispherical and radiating concretions of a grayish, white color, it occurs in the roofing-slate quarries at Peachbottom, in York county.

In very small globular, radiating, slightly reddish-white concretions, it is occasionally met with at the limonite mines near Chiques, Lancaster county.

128. *Coeruleo-lactite* (*Petersen*).

A pale blueish, greenish-blue or sky-blue mineral, which occurs in botryoidal incrustations upon a porous, silicious mineral, sometimes associated with small tufts of wavellite, is found at Gen. Trimble's limonite mines in East Whiteland township, Chester county. It is crypto-crystalline, and when magnified, of a vitreous lustre.

The coatings are usually so thin and mixed with other substances, that it is difficult to secure good material for analysis; that, which I had was not quite pure; enough, however, to determine the species.

I found:

			Coeruleo-lactite.
Sp. gr.	=	2.696	Calculated from analysis.
Phosphoric acid	=	36.31	37.04
Alumina	=	38.27	39.34
Cupric oxide	=	4.25	——
Insoluble	=	0.54	——
Water		21.70	23.62
		101.07	100.00

I give for comparison the analysis of Petersen of his coeruleo-lactite, from the Rindsberg Mine, near Katzenellnbogen, in Nassau, which it resembles more closely than any other mineral.

It is difficult to perceive in what combination the copper is present; that it belongs to this mineral is very probable, from the fact that none of the other minerals found at the same locality are colored by copper. Petersen's coeruleo-lactite contains 1.40 per cent. of oxide of copper.

129. *Cacoxenite* (*Steinmann*).

This hydrous ferric phosphate of doubtful composition, but supposed to be an iron-wavellite; occurs in yellow or brownish-yellow tufts or radiating masses.

Minerals corresponding with this description and containing

ferric oxide, phosphoric acid and water, have been found in the limonite mines at Conshohocken, Montgomery county; and with globular wavellite, at Chiques, Lancaster county.

Many of the limonites in this State (if not all) contain minute quantities of phosphoric acid. It is very probable that it is present as an admixture of a minute quantity of cacoxenite; while the phosphoric acid, which we so frequently find in magnetites, is present in the form of apatite.

130. *Torbernite* (*Werner*).

This mineral is found in very small quantity in Pennsylvania; it was, together with the following species, first observed by Mr. Theo. D. Rand, in the gneissic rocks opposite Fairmount, and subsequently at Deshong's quarry, near Leiperville, Delaware county. It occurs in minute tetragonal scales of a beautiful emerald-green color and pearly lustre.

It is a hydrous phosphate of uranium and copper.

131. *Antunite* (*Brooke and Miller*).

Crystallizes in the orthorhombic system, but usually occurs in nearly square, tabular crystals with eminent basal cleavage and pearly lustre. H. = 2–2.5; sp. gr. = 3.05.–3.19. Color sulphur- to citron-yellow. It is a hydrous phosphate of uranium and lime.

Beautiful square, pearly scales and crystalline crusts were formerly abundant in the gneiss, opposite Fairmount; inferior specimens were also obtained from Deshong's quarry, near Leiperville, and at Sam Crozer's quarry, near Upland, Delaware county, where it sometimes invests trapezohedral garnet.

Small scales of antunite have lately been found in a conglomerate from the neighborhood of Mauch Chunk; I have not succeeded however in learning the exact locality.

132. *Nitre.*

A mixture of nitrate of potash and lime has been frequently observed in the State, as an efflorescence or crystalline incrustation of old houses, &c.

V. *Tungstates and Molybdates.*

133. *Stolzite* (*Haidinger*).

Tungstate of lead or stolzite has been observed by Mr. Chas. M. Wheatley, at the Wheatley Mines, near Phœnixville, in small, well-defined, tetragonal octahedra of a yellowish-gray color. In some instances the stolzite crystals are enveloped by those of wulfenite.

134. *Wulfenite* (*Haidinger*).

It crystallizes in tetragonal forms, in square tables, modified by octahedra and prisms; sometimes massive, granular. Colors are waxy-yellow to orange-yellow, orange-red to bright hyacinth-red. H. = 2.75–3; sp. gr. = 6.03–7.01. Lustre adamantine to resinous. Its composition is molybdate of lead.

It has rarely been found in waxy-yellow, occasionally in bright orange, tetragonal tables with pyromorphite, &c., at the Ecton Mine, near Shannonville, Montgomery county;

In a great variety of forms, in tetragonal plates, obtuse and acute octahedra and cavernous crystals, of bright-yellow to deep red colors with many intermediate shades, at the Wheatley Mines near Phœnixville.

It has been analyzed by J. L. Smith (*l. c.*) who finds as follows:

		Yellow variety.	Red variety.
Sp. gr.	=	6.95	——
Molybdic acid	=	38.68	37.47
Vanadic acid	=	——	1.28
Plumbic oxide	=	60.48	60.30
		99.16	99.05

In grayish, yellowish-white tetragonal plates at the Pequea Mine, Lancaster county.

VI. *Sulphates, Chromates, Tellurates.*

α. *Anhydrous.*

135. *Barite.*

It crystallizes in orthorhombic prisms, usually modified by numerous domes and octahedra, and is usually found in crystals, also in crystalline columnar and radiating masses and granular; the cleavage is perfect parallel to the basal, and distinct, parallel to the prismatic planes. H. = 2.5–3.5; sp. gr.; = 4.3–4.7. Colors white, grayish, yellowish, blueish, brownish, brown. Lustre vitreous, inclining to resinous. Its composition is sulphate of baryta, BaO,SO_3 = baryta 65.7, sulphuric acid 34.3.

Very perfect transparent, greenish, tabular crystals of about half an inch in diameter, and clusters of blueish tabular crystals; also crested, fascicular and radiated crystals and crystalline masses, have been found at the Perkiomen Mine near Shannonville, Montgomery county.

In white laminated crystalline masses it occurs at the Phœnixville Mines; in a similar manner and also crystallized with copper ores, three miles west of New Hope, in Bucks county.

In the neighborhood of Brighton, in Beaver county, barite occurs in white laminated masses in nodules of argillaceous siderite.

Fibrous barite, associated with copper ores, is found at the Jug Hollow Mine, Montgomery county.

A fetid barite in brownish, radiating and columnar, ferruginous masses occurs at Heidelberg, Berks county; and a granular grayish-white barite, resembling marble, at Marble Hall, Montgomery county.

No analysis of Pennsylvania barite has been made.

136. *Celestite* (*Dana*).

Celestite or sulphate of strontia crystallizes in forms similar to those of barite, but in Pennsylvania only one variety has been observed, which is found at Bell's Mills, in Logan's Valley, near Frankstown, Huntingdon county.

It occurs in thin seams of a pale grayish-blue color and parallel-fibrous and columnar structure.

It has been analyzed already as early as 1797, by Klaproth (Beitraege, II, 92), who found:

Strontia	=	42
Sulphuric acid	=	58
		100

137. *Anglesite* (*Beudant*).

This mineral crystallizes orthorhombic, frequently in highly modified crystals, very similar to those of barite and celestite. The crystals are sometimes tabular, also often oblong, prismatic and acicular. It also occurs massive, but no massive varieties have been found in Pennsylvania.

H. = 2.75–3; sp. gr. = 6.12–6.39. It consists of sulphate of lead, PbO,SO_3, Plumbic oxide = 73.6; Sulphuric acid = 2.64

It has been found at the lead mines of Chester county,. especially Wheatley's, near Phœnixville, abundantly and in magnificent crystals of great variety; some of the crystals of nearly half a pound in weight were perfectly colorless; crystals terminated at both ends of five and a half inches in length and one and half an inches in thickness, have been obtained; perfectly limpid crystals of one inch were quite common. Sometimes it is colored black by an admixture of galenite, or delicate green by carbonate of copper, and yellow by oxide of iron.

The analysis by J. L. Smith (*l. c.*) shows it to be remarkably pure sulphate of lead. Colorless crystals of a sp. gr. = 6.35 contained:

Plumbic oxide	=	73.31	73.22
Sulphuric acid	=	26.78	26.61
Silica	=	0.20	——
		100.29	99.83

The anglesite crystals of Phœnixville are sometimes altered into cerussite.

Small but occasionally very fine crystals have been found at the Ecton Mine, near Shannonville, Montgomery county; and rarely at the Pequea Mine, Lancaster county.

138. *Vauquelinite* (*Berzelius*).

W. J. Taylor (Proc. Ac. Nat. Sc., Phila. 1858, 175) mentions minute crystals with acute terminations, in radiated aggregations, forming incrustations on quartz and galenite and of an apple- and siskin-green color. They have been found at the Pequea Mine, Lancaster county. He states that they gave chrome and copper reactions, and that he therefore considers them vauquelinite; but, as this mineral is a chromate of copper and lead, containing over sixty per cent. of the latter metal, and as Taylor does not speak of the *principal* constituent, there is a good deal of doubt about the correctness of his observation.

b. *Hydrous Sulphates.*

139. ? *Mirabilite* (*Haidinger*).

Mr. Theo. D. Rand (*l. c.*) states that in a cutting through decomposed mica schists, on the new line of the Philadelphia, Wilmington and Baltimore Railroad, about half a mile southwest of Gray's Ferry, he has observed white efflorescences, consisting chiefly of sulphate of soda.

Similar efflorescences he has found in the gneiss quarries near Chester, Delaware county.

140. *Gypsum.*

Crystallizes in monoclinic crystals, with an eminent clinodiagonal, and two other, less distinct cleavages; also foliated massive, often granular massive. H. = 1.5–2; sp. gr. = 2.314–2.328. Colorless, white, into yellow, flesh-red, gray, black, &c. Lustre subvitreous, on the clinodiagonal cleavage planes pearly. Consists of hydrous sulphate of lime. $CaO,SO_3 + 2H_2O =$ lime, 32.6, sulphuric acid, 46.5, water, 20.9.

Gypsum is a rare mineral in Pennsylvania. It occurs dissolved in some waters, especially *mine* waters; in the solid form I know it only in minute crystalline efflorescences, resulting from the action of oxydizing pyrite upon limestone, at Van Arsdale's, near Feisterville, Bucks county; and in beautiful slender crystals, sometimes two inches in length and less than one-eighth of an inch in width, but usually much smaller, at Cornwall. These minute, frequently needle-shaped, crystals occur upon magnetite, or upon a decomposed clay-like mineral, often intermixed with arborescent copper.

141. *Epsomite* (*Beudant*).

Epsom salt, the hydrous sulphate of magnesia, like some of the preceding minerals, occurs in Pennsylvania only in efflorescences, resulting from the action of pyrite upon magnesia minerals, mostly upon dolomite. In this manner it is found near Bethlehem and in the soapstone quarries near Lafayette, Montgomery county.

M. H. Boyé (Proc. Am. Phil. Soc. IV, 247) describes it as occurring in fibrous crystals and porous masses with bituminous coal on the Kiskiminetas River near Salzburg, Westmoreland county.

142. *Melanterite* (*Beudant*).

This name has been given to copperas or hydrous ferrous sulphate, when found as a mineral.

It occurs occasionally in stalactites at the Gap Mine, Lancaster county.

Formerly it was manufactured from an earth, found in limonite mines near Breunigsville, Lehigh county, and containing a large percentage of the same, resulting from the oxydation of marcasite or pyrite.

It frequently occurs in coal mines; and some of the mine waters near Ashland and elsewhere are so much contaminated with it that they are perfectly useless for domestic or steam-boiler purposes.

A simple and easy remedy for the purification of such waters would be of great importance; and I do not believe it would be a very difficult task to find a practical one.

143. *Goslarite* (*Haidinger*).

Hydrous sulphate of zinc or white vitriol has the mineralogical name goslarite. It is found as a result of the oxydation of zincblende in incrustations and fine needle-shaped, white crystals at Friedensville, Lehigh county.

It is a rare mineral.

144. *Bieberite* (*Haidinger*).

Only once I observed this rare mineral, a hydrous sulphate of cobalt, in minute quantity, as a flesh-colored crystalline incrustation upon magnetite, at Cornwall, Lebanon county.

145. *Morenosite* (*Casares*).

This hydrous sulphate of nickel is found usually mixed with other sulphates, as a greenish-white incrustation upon the nickel ores of the Gap Mine, Lancaster county.

146. *Calcanthite* (*v. Kobell*).

When the Gap Mine was first opened, the waters from it contained, according to some old records, a considerable quantity of sulphate of copper in solution. This has now almost entirely disappeared.

I do not know of any locality where blue vitriol has been observed as a mineral, except the soapstone quarries near Lafayette, Montgomery county, where it is occasionally met with as an efflorescence.

147. *Alunogen* (*Beudant*).

The hydrous sulphate of alumina is found in small, silky, fibrous incrustations in gneissic rocks, at Heft's mill, on the Wissahickon and at Hestonville, Philadelphia.

148. *Halotrichite* (*Glocker*).

Silky, fibrous, yellowish-white incrustations, consisting of a hydrous sulphate of alumina and ferrous oxide, are at some localities found in abundance; they are the result of decomposing pyrite upon the aluminous minerals in gneissic rocks.

It has been found in this manner, near Philadelphia, at Hestonville, and at the Columbia Bridge.

Also in South Coventry township, Chester county.

It probably occurs at many other places.

149. *Botryogen* (*Haidinger*).

This exceedingly rare hydrous ferroso-ferric sulphate, which was not known to occur at any locality except Fahlun in Sweden, has been observed by me in microscopic, globular, crystalline aggregations of a deep red color, associated with covellite and pyrite upon magnetite, at Cornwall, Lebanon county.

A qualitative examination corroborated my mineralogical determination.

150. *Glockerite* (*Naumann*).

It occurs in stalactitic brownish, resinous masses at the Columbia Bridge and Hestonville, Philadelphia; occasionally in considerable quantities at the Gap Mine, Lancaster county, resulting from the oxydation of the pyrrhotite.

It has been analyzed by Prof. W. Theo. Roepper, who kindly communicated the results of his analysis.

He found:

Ferric oxide	=	57.01
Sulphuric acid	=	13.97
Water	=	28.83
		99.81

This analysis corresponds with the formula: $2Fe_2O_3,SO_3 + 9H_2O$.

This is three atoms more water than usually adopted for glockerite, which has only six, but on account of the difficulty of obtaining it in a state of purity, this is no reason for considering it anything else.

151. *Brochantite* (*Levy*).

This combination of sulphate and hydrate of copper occurs at Cornwall, Lebanon county, where I have observed it in dark emerald- and blackish-green acicular crystals and crusts upon magnetite.

A qualitative analysis proved the correctness of my mineralogical determination.

152. ? *Langite* (*Maskelyne*).

There are associated with brochantite very small blueish-green crystals, containing sulphuric acid, oxide of copper and water, which are probably langite.

The crystals are too small and their quantity too minute for further investigation.

153. *Uraconite* (*Dana*).

Small quantities of so-called uranochre, a basic sulphate of uranium, have been found with other uranium minerals, in the vicinity of Chester, Delaware county.

VII. *Carbonates.*

a. *Anhydrous.*

154. *Calcite* (*Haidinger*).

Calcite occurs in Pennsylvania in many forms, both crystallized and in granular and compact varieties, the latter constituting the numerous beds of marble and limestone, which abound in all the geological formations of the State. As they will be dwelt upon more fully in the geological reports, I confine myself principally to such localities, where crystallized varieties and unusual forms have been observed.

Calcite crystallizes in the hexagonal system with a rhombohedron of 105° 5′ as the fundamental form, in combinations with about 50 additional rhombohedra, about 90 scalenohedra, ten pyramids, &c., forming innumerable varieties of crystals. The cleavage of calcite is very eminently parallel to the fundamental rhombohedron, which is observable in the smallest granular varieties, giving them a lustrous appearance. It is also found coarse and fine fibrous and columnar, sometimes lamellar, and, as already mentioned, granular, compact and earthy. Frequently in stalactitic, tuberose, nodular and imitative forms. H. = 2.5–3 ; sp. gr. = 2.5–2.78. Lustre vitreous ; colorless, white, shading through gray, yellow, red, violet, brown into black. The colorless crystallized varieties, so-called "Iceland spar," show strong double refraction. It consists, when pure, of carbonate of lime = CaO,CO_2, containing : lime, 56, carbonic acid, 44, but often contaminated with silicic acid, alumina, and especially with carbonates of magnesia, manganese, iron, &c. ; and frequently graduating into dolomite, breunnerite, &c.

Many beautiful crystals have been found in the State, but their combinations have been studied but very little.

At Bullock's quarry near Conshohocken, Montgomery county, very fine and perfect crystals have been observed, mostly combinations of hexagonal prism, scalenohedra and rhombohedra.

Large cleavage crystals, both of the rhombohedron and hemitrops are found in Jacoby's quarry near Conshohocken; at Hitners' quarry at Marble Hall, we find several varieties of calcite, the most important of all, a white granular marble, largely in use in the City of Philadelphia, and elsewhere, for building and ornamental purposes; besides this, scalenohedra with rhombohedral terminations have been found, and large cleavage crystals; also a pink variety with curved planes.

Hexagonal prisms, also massive calcite and purple and flesh-colored, cleavable calcite are found at Easton.

Very remarkable crystals of calcite have been found at the Wheatley mines near Phœnixville. Sometimes slabs have been found with a surface of 8 to 10 square feet, covered with prismatic crystals an inch or two in length, and from one-quarter to one inch in thickness. These crystals are occasionally 8 or 10 inches in length, only a quarter of an inch in diameter, preserving their hexagonal shape throughout their entire length.

This mine furnishes also very remarkable aggregations of crystals, arranged around an axis, so as to form a double spiral; others containing scalenohedral crystals with a small cubic crystal of fluorite on the summit of each, inclosed in crystals, which are combinations of hexagonal prisms and rhombohedra; others again with pyrite symmetrically arranged in the rhombohedral planes of the summit. (J. L. Smith, *l. c.*)

At the tunnel on the Reading Railroad, at Phœnixville, good crystals of calcite have been found associated with pyrite; they are obtuse rhombohedra with a hexagonal prism.

Scalenohedral crystals are found at the Poorhouse quarry, and obtuse rhombohedra and fibrous calcite in the limestones at Downingtown, and good cleavage masses in the limestone quarries of Birmingham and London Grove townships, and, inclosing byssolite, in the iron mines near Knauertown, Chester county.

Fine specimens of calcite have been found in Monroe county and Iceland spar in York county.

It has been found in very obtuse rhombohedral crystals upon chromite at Wood's Mine, and at Reynold's Mine, Lancaster county.

Calcite has also been met with at Frankford, the Columbia Bridge and Flat Rock Tunnel, near Philadelphia.

At Fritz's Island, near Reading, it is met with in crystals and granular; it is also found at Jones' Mine in Berks county. Also in globular concretions and incrustations colored pink by cobalt upon magnetite at Cornwall, Lebanon county.

It would be tedious to mention all the localities, where calcite has occurred in limestone beds throughout the State, in cleavage masses or crystalline forms.

Stalactitic calcite in various shapes is found in the limonite mines of West Whiteland township, Chester county, and in hollow quills at Catasauqua, Lehigh county.

Most interesting are the stalactites of the Crystal Cave near Kutztown, in Berks county, which are partly calcite, but also frequently aragonite, often on the same piece, some of the stalactites being terminated with rhombohedra, and studded at the same time with crystals of aragonite.

The very soft, white, earthy calcite, usually known as *rock-milk* or *agaric mineral*, is frequently met with in the lower Silurian limestones near Easton, and elsewhere.

A most beautiful, and when polished, jet black marble occurs in the Mosquito Valley, Armstrong township, Lycoming county.

155. *Dolomite* (*Kirwan*).

It crystallizes like calcite in rhombohedral crystals, but with angles of 106° 15′. The crystals are sometimes modified by hexagonal prisms and rhombohedra, frequently with curved faces, also scalenohedra, usually hemihedral. The cleavage is perfect parallel to the fundamental rhombohedron. It is also found in coarse and fine granular masses. H. = 3.5–4; sp. gr. = 2.8–2.9. Color white, also reddish and grayish-white, gray, brown, black. Lustre vitreous, inclining to pearly. The composition of the typical dolomite is $CaO,CO_2 + MgO,CO_2$, containing 54.35 per cent. of carbonate of lime and 45.65 per cent. of carbonate of magnesia; but there are innumerable gradations in both directions, into calcite on the one side, and magnesite on the other.

Very large beds of dolomite or magnesian limestones are found in the State; they form, for the most part, the lower strata of the Silurian limestones.

I will not repeat the many analyses, already published in H. D. Roger's Report, and will only for comparison give a few analyses, which I have made and published in my "Investigation of iron ores and limestones from Messrs. Lyon, Shorb & Co.'s iron ore banks, on Spruce Creek, Halfmoon Run and Warrior's Mark Run, in Centre, Blair and Huntingdon counties, Pa." (Proc. Am. Phil. Soc., XIV, 84–96), to which I add the analysis of a limestone from one mile north of Alburtis, on the road to Fogelsville, Lehigh county, which Mr. H. Pemberton, Jr., made at my request; and a few determinations of the amount of carbonate of magnesia in limestones, analyzed by Prof. W. Th. Roepper, the results of which he kindly communicated to me:

		1. Fine crystalline granular whitish and gray limestone at head of Hostler Bank.	2. Very fine granular ash-gray limestone in Hostler Bank.	3. Slightly crystalline, dark-gray, compact upper limestone from Penna. Bank.	4. Pale ash-gray, very fine crystalline limestones in Pennsylvania Bank.	5. Yellowish-gray, soft, rotten limestones in Pennsylvania Bank.	6. Ash-gray, finely granular, from near Alburtis, by H. Pemberton, Jr.
Carbonate of lime	=	59.44	51.82	72.67	51.25	45.73	48.29
" magnesia	=	35.19	42.52	3.98	42.39	35.51	37.74
" manganese	=	0.19	0.24	0.18	0.06	trace	——
" iron	=	0.80	0.50	1.31	0.45	1.18	0.61
Quartz and Silicic acid	=	3.84	4.33	18.05	5.03	15.83	12.13
Alumina	=	0.54	0.42	3.81	0.82	1.75	1.18
Water	=	——	0.17	——	——	——	——
		100.00	100.00	100.00	100.00	100.00	99.95

Roepper's analyses gave:

Fossiliferous limestone from Solomon's Gap, about four miles from Wilkes-barre,	1.51 per cent.	carbonate of	magnesia.
Slaty variety from same locality,	4.19	"	"	"
Limestone from North White Hall,	3.60	"	"	"
Black subcrystalline from Mori's quarry, Saucon Valley,	4.74	"	"	"
South White Hall limestone (16 per cent. insoluble),	29.52	"	"	"

Limestone from Hellertown (11.82 per cent. insoluble),	36.46 per cent. carbonate of magnesia
Blueish-gray crystalline, same locality 2.81 per cent. insoluble),	36.58 " " "
Grayish-black crystalline, from Saucon Valley,	39.99 " " "

Crystallized dolomite is frequently met with in fractures or fissures in the dolomitic limestone; for instance, at Easton; New Britain, Bucks county; Avondale and Poorhouse quarries, and a fetid variety at Caleb Cope's quarries, East Bradford township, Chester county.

The so-called pearlspar, in very fine slightly curved rhombohedral crystals, has been found at the Phœnixville tunnel.

Small but brilliant crystals, which are combinations of hexagonal prisms with a rhombohedron, with curved faces, and good cleavable masses of a grayish-white color, occur with talc at the soapstone quarries at Lafayette, Montgomery county.

Cleavable masses, usually mixed with the fibrous serpentine variety "baltimorite," are found at Wood's Mine, Lancaster county.

At the same locality are found peculiar grayish-white incrustations, frequently colored green by a small admixture of zaratite, which are usually called "pennite." I have analyzed a carefully selected sample of these grayish-white incrustations, the individual particles of which, when examined under a strong lens, appeared to be hexagonal prisms, terminated with an obtuse rhombehedron, and found the composition to be:

Carbonate of lime	=	52.64
Carbonate of magnesia	=	46.83
		99.47

which is the composition of a true dolomite.

153. *Ankerite (Haidinger).*

A dolomite, in which a large, but variable portion of the magnesia is replaced by ferrous and sometimes by manganous oxide, has been called ankerite. Its forms are similar to dolomite; usually in crystalline masses of a coarse and fine granular structure; also compact. Color white, yellowish-white, gray, frequently, by oxydation, changing to brown.

It is probable that many varieties of the so-called brownspar or breunnerite belong to this species, a question which cannot be decided in the absence of analyses.

It occurs at the Phœnixville Mines in curved rhombohedral crystals, but usually in yellowish-white, crystalline, granular masses.

It has, at my request, been analyzed in the Laboratory of the University of Pennsylvania, by Dr. W. P. Headden, who found:

Carbonate of lime	=	50.72
" " magnesia	=	21.98
" " iron	=	27.29
Insoluble	=	0.22
		100.21

157. *Magnesite* (*Karsten*).

Crystallizes in rhombohedral crystals; also, massive, fibrous, granular to compact. Cleavage very perfect, parallel to a rhombóhedron with an angle of 107° 29′. H. = 3.5–4.5; sp. gr. = 2.8–3.2. Colorless, white, grayish and yellowish-white, brown. Its composition is carbonate of magnesia = MgO,CO_2, containing, when pure, 47.6 per cent. of magnesia and 52.4 per cent. of carbonic acid.

Besides the pure variety, there is one, more abundant than the first, which has from $\frac{1}{25}$ to $\frac{1}{4}$ of the magnesia replaced by ferrous oxide. This ferriferous variety is usually called breunnerite (Haidinger) or brownspar.

Cleavage masses of magnesite, containing but a small percentage of ferrous oxide, occur rarely at Wood's Mine, and the adjoining Carter's Mine, in Lancaster county.

Radiating crystals of magnesite are found at R. Taylor's mill, and one mile north of West Chester, in West Goshen township; in Brinton's quarry, three miles south of West Chester, in Westtown township, occurs a small seam of yellowish-white earthy material in serpentine, which Mr. W. W. Jefferis considers to be magnesite; it is also found at Strode's mill, East Bradford township, Chester county.

Massive, compact, white magnesite is found in Lancaster county at Low's Chrome Mine; also near Scott's Pit, and abundantly at the magnesia quarries, on Goat Hill, or, as it is sometimes called, Mount Ararat, in West Nottingham township, Chester county.

A specimen which has been analyzed at my request, in the Laboratory of the University of Pennsylvania, by Mr. John H. Campbell, contained:

Silicic acid	=	3.50
Carbonic acid	=	47.97
Alumina	=	0.45
Ferrous oxide	=	trace
Magnesia	=	45.96
Lime	=	0.40
Water	=	2.46
		100.56

This analysis shows the presence of 91.58 per cent. of pure carbonate of magnesia, and an admixture of 8.42 per cent. of impurities, mostly serpentine.

It has also been found in Radnor township, Delaware county, and in Spang's old iron mine, Spangsville, Earl township, Berks county.

The ferriferous variety, *breunnerite*, occurs in fine rhombohedral crystals in the chlorite slates of the soapstone quarries, and in small crystals and crystalline masses, in the Perkiomen Mines, Montgomery county.

It has also been observed at Easton.

The white, compact magnesite was formerly mined to a considerable extent at the magnesia quarries.

It occurs in the serpentine, in irregular veins, from a few inches sometimes to ten feet, of which, however, only about two and a half feet were pure.

From the information, which I obtained from Mr. W. Wallace Wiley, an old resident of the neighborhood, who had been employed at the mines, since they were first opened in 1835, I learn that between 1835 and 1840, about one hundred and fifty tons per annum were shipped to Messrs. Handy and John Ellicott, of Baltimore; between 1840 and 1850, on an average also about one hundred and fifty tons a year, to Messrs. Samuel and Philip Ellicott. The mines were then dormant for four years, when they were re-opened in 1854, and worked until 1871, by Messrs. Powers and Weightman, of this city, producing annually an average of five hundred tons of carbonate of magnesia. They are not in operation at present.

This quantity of carbonate of magnesia was used in the manufacture of epsom salt, one ton yielding two and three-quarter tons of epsom salt.

The discontinuation of the mining operations at the magnesia quarries was caused by the discovery of large masses of kieserite (a sulphate of magnesia with one atom of water) in the rocksalt mines of Stassfurt in Prussia, which on boiling takes up a large quantity of water and is converted into epsom salt.

158. *Siderite* (*Haidinger*).

Crystallizes in rhombohedral crystals with an angle of 107°. Besides the fundamental, a few other rhombohedra occur, together with scalenohedra, hexagonal prisms and the basal plane. The crystals have often curved faces. The cleavage parallel to the fundamental rhombohedron is perfect; it is also found in botryoidal, fibrous forms or in cleavable masses; coarse and fine granular, compact. H. = 3.3–4.5; sp. gr. = 3.7-3.9. Color yellowish-white, yellowish-gray, ash-gray, brown.

It consists, when pure, of carbonate of iron = FeO,CO_2, containing 62.1 per cent. of ferrous oxide and 37 9 per cent. of carbonic acid.

Frequently it is contaminated with a considerable quantity of clay, forming nodular masses of the so-called argillaceous or clay iron ore; or with carbonaceous matter as the so-called blackband iron ore.

The immediate connection which has been observed at several localities in Pennsylvania and lately in Delaware, between the siderite and limonite, leaves very little doubt that many of the limonite beds in various formations are the result of the decomposition of the siderite, involving the loss of its carbonic acid, the oxydation of the ferrous into ferric oxide and the hydration of the latter.

This alteration occurs in limonite beds of the Lower Silurian limestone at Ironton, in Lehigh county, and near Hellertown, Northampton county, and on V. Hartman's farm, near Reading, in Berks county, where the siderite is found in the lower part of the beds.

It has usually a pale-brownish, grayish-white color and a granular structure, and the cavities sometimes contain small crystals. A specimen from Wurst's Mine, at Hellertown, has been analyzed by Prof. W. Th. Roepper (priv. com.) (see below).

Although not within the State, I will on account of its interest, mention here the discovery of a similar granular, grayish-white siderite in the lower parts of the large limonite deposit at Newark, Delaware.

Of very great importance are the occurrences of siderite in the lower part of limonite beds, found in Woodcock Valley, west of Broadtop and in Aughwick and Juniata Valleys, between Orbisonia and Lewisburg, in Mifflin county.

A specimen from the Ross ore opening at McVeytown Gap, Mifflin county, has, at my request, been analyzed in the Laboratory of the University of Pennsylvania, by Mr. Pedro G. Salom (see below).

Small rhombohedral crystals and combinations of a flat rhombohedron with a hexagonal prism, and crystalline incrustations of siderite are found at the Gap Mine, Lancaster county.

Small brown rhombohedral crystals and globular aggregations with kaolinite upon quartz, near Pottsville, Schuylkill county.

Lenticular grayish-brown crystals are found in the cavities of kidney-shaped and globular argillaceous iron ore from the carboniferous shales of Scranton, Luzerne county.

Granular grayish and brownish-white siderite is occasionally met with in Deshong's gneiss quarries, near Leiperville, Delaware county.

The argillaceous or clay iron ores are of great importance in Pennsylvania.

As their geological position and value, as ores, will be more fully investigated and discussed by other gentlemen connected with the survey, I will merely give the limits of their occurrence and add a single analysis, which I made of a compact gray variety from Gillhausen's Vein, Indian Camp, Jefferson county.

The so-called Ralston ore bed is found throughout the whole extent of the coal measures, underneath the conglomerate.

In the lower coal measures clay iron ores, more or less altered into limonite, occur in Venango Armstrong, Butler, Jefferson, Lawrence, Beaver and Cambria counties; they are also found under and over the Pittsburg coal bed, in Westmoreland and Fayette counties.

The following are the results of the analyses:

		1. Wurst's Mine, Hellertown. ROEPPER.	2. McVeytown Gap. SALOM.	3. Gillhausen's Mine, Indian Camp. GENTH.
Carbonate of iron	=	90.51	82.05	78.34
" manganese	=	0.78	1.41	0.54
" cobalt	=	——	0.27	——
" magnesia	=	1.20	0.71	7.39
" lime	=	——	4.39	6.07
Alumina	=	——	7.02	0.38
Silicic acid, &c.	=	8.00	5.06	6.93
Organic matter	=	——	0.40	0.35
		100.49	101.31	100.00

159. *Rhodochrosite* (*Hausmann*).

A calciferous variety of carbonate of manganese or rhodochrosite, is occasionally met with at Cornwall, Lebanon county, where it occurs in globular concretions in cobaltiferous wad.

160. *Smithsonite* (*Beudant*).

It crystallizes in small rhombohedral crystals with perfect cleavage parallel to the planes of a fundamental rhombohedron of 107° 40′. Several other rhombohedra have been observed together with one scalenohedron, a hexagonal prism, and the basal plane; also in reniform, botryoidal, stalactitic masses, and in crystalline incrustations; granular. H = 5; sp. gr. = 4–4.45. White, grayish and brownish-white; lustre vitreous, inclining to pearly. When pure, consists of carbonate of zinc = ZnO,CO_2, containing 64.8 per cent. zincoxide and 35.2 per cent. carbonic acid.

It is found in small, but brilliant, white or grayish-white scalenohedra and in granular and globular masses at Friedensville, Lehigh county.

In granular masses, sometimes showing separation in laminæ, in Sinking Valley, Blair county.

Granular brownish-gray and in pseudomorphs after dolomite, at the Lancaster Zinc Mine.

Carbonate of zinc or smithsonite is a very valuable zinc ore.

161. *Aragonite* (*Werner*).

It crystallizes in orthorhombic prisms, usually modified by domes and octahedra; frequently in twins; also columnar, with straight or divergent fibres, incrusting, stalactitic, globular, reniform or coralloidal. It has no perfect cleavages.

H. = 3.5–4; sp. gr. = 2.927–2.947. White, yellowish, grayish, &c. Lustre vitreous, sometimes inclining to resinous; the fibrous varieties sometimes silky. It is like calcite composed of carbonate of lime, CaO,CO_2 containing lime 56 per cent. and carbonic acid 44 per cent.

Most beautiful groups of radiated crystals have been found at Wood's Mine, Lancaster county.

In acicular crystallizations and botryoidal and fibrous coatings, it is found at Jones' and sparingly at Fritz's Island Mines, Berks county.

In the same county it is found at Kutztown, in the Crystal Cave, abundantly in small crystals, forming crystalline radiating masses; also in fine fibrous stalactites radiating from a centre; some of the stalactites being in part composed of calcite and in part of aragonite.

In Chester county the fibrous and stalactitic aragonite occurs in the iron mines near Knauertown; groups of crystals, the latter from one to one and a half inches long, have been observed at the Wheatley Mines.

In columnar, fibrous, sometimes radiating seams, it occurs in the serpentine at R. Taylor's Mill, West Goshen township; in fibrous and radiating seams of about one-half an inch thick at Brinton's quarry, in Westtown township; it is also found in Nivin's quarry, New Garden township; in London Grove and Kennett townships, Chester county.

In silky, fibrous, crystalline seams and crusts it is found on the Ohio River, at Pittsburg.

A highly interesting *zinciferous* variety of aragonite, in crystals of three-fourths to one inch in length, forming radiating groups of a white and yellowish-white color, has been observed at the Friedensville Zinc Mines, by Prof. W. Theo. Roepper, who kindly furnished me with his analysis.

It contains:

Carbonate of lime	=	94.20
" zinc	=	4.73
Insoluble	=	0.53
		99.46

Most of the aragonite from Friedensville contains carbonate of zinc.

162. *Cerussite* (*Haidinger*).

It crystallizes in orthorhombic crystals, often highly modified; frequently in twins, also granular massive and compact. H. = 3.35; sp. gr. = 6.47-6.48. Colorless, yellow, grayish-white, brownish, grayish-black. Lustre adamantine, inclining to vitreous and resinous. Consists of carbonate of lead = PbO,CO_2 containing plumbic oxide 83.5, carbonic acid 16.5.

In large and beautiful colorless, white and wine-yellow crystals at the Phœnixville lead mines, especially the Chester county and Wheatley's; ochre-yellow massive. A black variety has also been observed.

Very interesting are the pseudomorphs after anglesite, which form mere shells, generally a thin coating of limonite representing the original form, the cavity studded with crystals of cerussite.

Cerussite in minute crystals occurs as a coating upon galenite at the Pequea Mine and the Lancaster Zinc Mines, Lancaster county.

It occurs massive with smithsonite in Sinking Valley, Blair county; and in black and white crystals and crystalline masses at the Ecton Mine, near Shannonville, Montgomery county.

J. L. Smith (*l. c.*) published the analysis of a colorless crystal from the Wheatley Mine near Phœnixville.

It had a spec. gr. of 6.60, and contained:

Lead oxide	=	83.76
Carbonic acid	=	16.38
		100.14

The upper levels of the Chester county Silver-lead Mines and the Ecton Mine, Montgomery county, have furnished considerable quantities of this valuable lead ore.

b. *Hydrous Carbonates.*

163. *Hydromagnesite* (*v. Kobell*).

Crystallizes in small acicular and bladed crystals or tufts, also amorphous, as chalky or earthy crusts. H. = 3.5 ; sp. gr = 2.18. Color white ; lustre vitreous, inclining to silky or pearly.

It is a very rare mineral, found at Low's Mine and Wood's Mine, Lancaster county, in radiated crystallizations or in globular, radiating masses or small crystals, implanted in brucite.

The mixture of hydromagnesite and brucite was formerly considered a distinct species and was called "lancasterite."

It has been analyzed by Smith and Brush (Am. Journn. Sc. [2] XV, 214), who found:

		Low's Mine.	Wood's Mine.
Carbonic acid	=	36.74	36.69
Magnesia	=	42.30	43.20
Water	=	20.10	19.43
Iron and Manganese		trace	trace
		99.14	99.32

164. ? *Hydrodolomite* (*Rammelsberg*).

I have not been able to satisfy myself of the existence of this species at Wood's Mine, Lancaster county ; some of the specimens, which are usually called *pennite*, are certainly dolomite, as I have shown by the analysis given above ; it will require further investigations to ascertain whether Hermann's view is correct.

The analysis which he gives of his "pennite" from Texas (Journ. Prakt. Chem. XLVI, 33) is as follows:

Lime	=	20.10
Magnesia	=	27.02
Ferrous oxide	=	0.70
Nicolous oxide	=	1.25
Manganous oxide	=	0.40
Alumina	=	0.15
Carbonic acid	=	44.54
Water	=	5.84
		100.00

165. *Lanthanite* (*Haidinger*).

A single specimen of one of the rarest minerals, well crystallized carbonate of lanthana, was found about twenty-two years ago near Bethlehem. The exact locality is not known; it has been asserted that it came from the limestones at the zinc mines, but it is more probable that it came from the neighborhood of the allanite locality, and is the result of the decomposition of allanite.

Under "allanite" it has been stated that this is always coated with a crust of altered mineral; it is probable, therefore, that the lanthana from it separated and crystallized in the form of a hydrous carbonate.

The original specimen was hardly of the size of a man's fist, and consisted of an aggregate of small orthorhombic crystals in thin four-sided plates or minute tables, with beveled edges and a micaceous cleavage; it shows pearly lustre, and a delicate pinkish, grayish-white color. H. = 2.5–3; sp. gr. = 2.605–2.666.

It was first described and its forms figured by W. P. Blake (Am. Journ. Sc. [2] XVI, 228), who gave several analyses, closely agreeing with the later analyses by J. L. Smith (Am. Journ. Sc. [2] XVI, 230, and XVIII, 378), and by myself (Am. Journ. Sc. [2] XXIII, 425):

		Smith.		Genth.
Lanthana (and Didymia)	=	54.90	55.03	54.95
Carbonic acid	=	22.58	21.95	21.08
Water	=	24.09	24.21	[23.97]
		101.57	101.19	100.00

166. *Zaratite* (*Casares*).

This beautiful mineral, from Wood's Mine, near Texas, Lancaster county, was first noticed by B. Silliman, Jr. (Am. Journ. Sc. [2] III, 407), who considered it, before a complete analysis was made, a hydrate of nickel. He subsequently found that it was a hydrous carbonate and called it "emerald-nickel" (Am. Journ. Sc. [2] VI, 248). It was subsequently analyzed by Smith and Brush (Am. Journ. Sc. [2] XVI, 52).

Amorphous; often small stalactitic and minute mammillary,

massive, compact. H. = 3–3.25 ; sp. gr. = 2.57–2.693 ; emerald-green ; lustre vitreous. Fracture subconchoidal ; brittle.

The analysis gave :

		Silliman.	Smith and Brush.
Nickel oxide	=	58.81	56.82
Magnesia	=	——	1.68
Carbonic acid	=	11.69	11.63
Water	=	29.50	29.87
		100.00	100.00

agreeing with the formula $NiO,CO_2 + 2\ NiO,H_2O + 4\ H_2O$.

It has also been found in small quantities at the Red Pit and Low's Mine, Lancaster county, and Moro Phillip's Chrome Mine, West Nottingham, Chester county.

167. *Hydrozincite* (*Kenngott*).

Massive, earthy, compact. In mammillated, stalactitic incrustations, sometimes concentric ; also earthy or chalk-like.

H. = 2–2.5 ; sp. gr. = 3.58–3.8. It is a combination of carbonate and hydrate of zinc = $ZnO,CO_2 + 2\ ZnO,H_2O$.

Fine specimens have been found at Friedensville, Lehigh county, consisting of mammillated incrustations or porcelain-like and earthy masses.

It has rarely been met with in white incrustations upon smithsonite at the Lancaster Zinc Mines, Lancaster county.

168. *Aurichalcite* (*Böttger*).

It is a carbonate and hydrate of zinc and copper.

Sparingly found at the Lancaster Zinc Mine in fine pale blueish-green scales or incrustations upon dolomite or smithsonite.

169. *Malachite* (*Wallerius*).

Crystallizes rarely in distinct monoclinic forms, usually massive and in tuberose, botryoidal and stalactitic incrustations with divergent structure ; often in delicate tufts of fibrous crystals or compact fibrous masses, banded in color, also granular and earthy.

H. = 3.5–4 ; sp. gr. = 3.7–4. Color, bright emerald to grass-green, with adamantine lustre, inclining to silky for the fibrous varieties, also dull and earthy. It is a combination of carbonate of copper with hydrate of copper=CuO,CO_2+CuO,H_2O, containing copper oxide 71.9, carbonic acid 19.9, water 8.2.

Beautiful, very fine fibrous and botryoidal masses have been found at Jones' Mine, near Morgantown, Berks county; some of the specimens are similar to the celebrated Siberian malachite and admit of a high polish.

Fibrous malachite occurs also at Fritz's Mine, near Reading; very good fibrous and compact varieties at Cornwall, Lebanon county.

Similar varieties are found at the mines near Knauertown, in Warwick township, Chester county; Nicholson's Gap, and at the silver-lead mines, and the copper mines near Phœnixville.

It has been found in good specimens at Perkiomen Mine, near Shannonville; it also occurs occasionally at Easton and three miles west of New Hope and in New Britain, Bucks county.

In small quantity it is occasionally met with at Frankford, at McKinney's quarry and on corundum at Unionville.

Associated with chalcocite it stains the rocks at many places along the South Mountains and in the copper horizon of the Catskill group in Northern Pennsylvania; and frequently the copper bearing epidotic rocks of Adams county.

A fibrous malachite of Phœnixville, of 4.06 sp. gr., has been analyzed by J. L. Smith (*l. c.*):

Cupric oxide	=	71.46
Carbonic acid	=	19.09
Water	=	9.02
Ferric oxide	=	0.12
		99.69

170. *Azurite* (*Beudant.*)

Crystallizes in monoclinic crystals, frequently highly modified, also massive with columnar and radiating structure, granular, earthy. H. = 3.5–4 25; sp. gr. = 4.5–3.8. Various shades of azure blue. Lustre, vitreous. Consists of two atoms of carbonate of copper, with one atom of hydrate of copper = $2CuO,CO_2 + CuO,H_2O$, containing 69.2 per cent. copper oxide, 25.6 carbonic acid and 5.2 water.

It occurs in very fine crystals, sometimes over a quarter of an inch in size and groups of crystals or crystalline crusts, usually of dark, but also of pale azure-blue color, at Cornwall, Lebanon county.

Sparingly in small crystals, at Fritz's Island, near Reading.

Good crystals occur with wulfenite in and on chalcopyrite at the Perkiomen Mine, near Shannonville, Montgomery county.

Very good, deep azure-blue crystals of a quarter to a half inch across, are found on chalcopyrite, at the Wheatley and other mines near Phœnixville.

The latter have been analyzed by J. L. Smith (*l. c.*), who found the sp. gr. = 3.88, and the composition:

Cupric oxide	=	69.41
Carbonic acid	=	24.98
Water	=	5.84
		100.23

171. *Bismutite* (*Breithaupt*).

Very small quantities of this hydrous carbonate of bismuth have been found by Mr. Th. D. Rand, in the gneissic rocks of the vicinity of Chester, Delaware county.

APPENDIX.

CHAPTER VI.

HYDROCARBON COMPOUNDS.

Minerals of Organic Origin.

In Organic Chemistry we find a large class of compounds, consisting of simply the two elements, carbon and hydrogen, and known under this name of *Hydrocarbons*. They are very important as constituting the starting-points, from which innumerable other classes, such as the Alcohols, the Phenols, and the Organic Acids proceed.

These Hydrocarbons, we find, can be arranged in a number of series, each member of which will be found in its composition to answer some simple general formula and to differ from the preceding member of its series by the constant increment CH_2. These are known by Chemists as *homologous series*. Chemical theory allows of an indefinite number of such series of Hydrocarbons. Practically we know of some ten series of this kind, several of them quite numerous.

The Pennsylvania Petroleum is proved to be a mixture of such Hydrocarbons, containing members of several of these series. By long-continued fractional distillation, they can be isolated in a state of purity, and their identity with the compounds formed in the Laboratory proven. Indeed many members in some of these series were first obtained from the Petroleum, and have not as yet been produced artificially.

The origin of Petroleum is unknown. By some it has been attributed to the decomposition of vegetable substances alone. This view was held by Bischof (Chem. Geol. I, 789, 1863).

Reichenbach even goes so far as to consider the Petroleum as the turpentine of the pines of the corresponding geological period. It is, however, now generally admitted that it has come from animal as well as vegetable remains, as urged by Dufrénoy, Newberry and others.

The similarity of the Petroleum to the Oils obtained by the destructive distillation of Cannel Coals and Bituminous Shales is most striking. There are important differences however, and certain industries based upon ingredients of the latter do not appear likely to connect themselves with the former.

Petroleum as an obvious mixture and a very varying one, too, cannot be assigned a place as a mineral species. As Dana properly points out (Mineralogy, 1868, 722), such mixtures partake more of the nature of rocks than minerals.

However we can enumerate the different distinct Hydrocarbons that have been definitely ascertained as occurring in it, and refer them to the proper series of *homologous* compounds.

A. *Marsh-gas Series* (*Paraffines*).

General formula—C_nH_{2n+2}

172. *Methyl Hydrid* (Marsh-gas), CH_4.

Is a colorless, odorless and inflammable gas of 0.559, sp. gr. Is not condensable. With air forms the well-known "firedamp" of coal mines.

Was found by Fouqué (Compt. rend. LXVIII, 1045) to be present in the gaseous exhalations from petroleum wells at Petrolia (and Fredonia, N. Y.).

173. *Ethyl Hydrid* (*Aethan*), C_2H_6.

Is a colorless, almost odorless, gas. Inflammable, with slightly luminous flame.

Was found also by Fouqué (*loc. cit.*) in the same exhalations at Petrolia (and Fredonia).

Exists, moreover, dissolved in the crude petroleum, from which it was collected and analyzed volumetrically by Ronalds (Jour. Chem. Soc. [2] vol. 3, 54).

174. *Propyl Hydrid* (*Propan*), C_3H_8.

Is a colorless gas, which is liquid under —17° C.

Was found by Fouqué (*loc. cit.*) in the gaseous exhalations from petroleum wells at Pioneer Run.

Was collected from crude petroleum and analyzed volumetrically by Ronalds (*loc. cit*).

Lefebvre (Compt. rend. LXVII, 1353), by condensing the vapors with freezing mixtures, obtained from crude petroleum a compound, which he calculated to have a boiling point of —25° to —30° C, and which he supposes was C_3H_8.

175. *Butyl Hydrid* (*Butan*), C_4H_{10}.

There exist two compounds, both of which possess this formula, although their molecular structure is obviously different. They are what Organic Chemistry calls *Isomeric compounds.* Both appear to exist in the petroleum.

a. *Normal Butan.*

This is a colorless gas condensing at 1° C.

It has been isolated in a pure state by Warren (Am. Jour. Sc. [2] XL, 226) who determines its boiling point as 0° C.

Lefebvre (*loc. cit.*), in the manner already stated, obtained evidence of a compound of this composition, to which he assigns a boiling point of 0°, although he did not isolate it.

b. *Pseudo-butan.*

Is a colorless gas, condensing to a liquid at 17° C. This does not appear to have been obtained as yet in a pure state.

Warren (*loc. cit.*) finds an isomeric body of this composition boiling at 8° to 9°. There are, however, only the two isomers possible, so this must have been a mixture of the two butans.

Pelouze and Cahours (Comp. rend. LVI, 505) speak of a portion of the petroleum as boiling at 0°, although they did not analyze it.

Schorlemmer (Ann. der Ch. und Ph. CXXVII, 313) speaks of the lightest portions of the petroleum as boiling at 20° to 30°, and probably containing C_4H_{10}.

Ronalds (*loc. cit.*) found a portion having the composition C_4H_{10} to have the boiling point 4° to 6°, and determined its sp. gr. at 0° as 0.600. This was probably a mixture like Warren's second compound.

Fouqué (*loc. cit.*) found in the gas exhaled at Pioneer Run evidence of C_4H_{10}.

176. *Amyl Hydrid* (*Pentan*), C_5H_{12}.

There are three isomeric bodies of this composition. Two of them appear to be contained in the Pennsylvania petroleum.

a. *Normal Pentan.*

Is a colorless, very mobile liquid boiling at 37°–39° C. This has been obtained in a pure state, by Warren (*loc. cit*), and Schorlemmer (Ann. der Ch. und Ph., CLXI, 268).

b. *Dimethyl-propan.*

Is a colorless liquid, boiling at 30° C., which does not solidify at —24° C. Its sp. gr. is 0.626 at 17° C.

This has been obtained in a pure state, by Pelouze and Cahours (*loc. cit.*), by Warren (*loc. cit.*), who makes its boiling point 30°.2 C., by Schorlemmer (Ann. Ch. und Ph., CXXXVI, 268). This last chemist had previously determined the boiling point of Pentan as 35° C., but acknowledges in his last article that he had worked with a mixture.

177. *Hexyl Hydrid* (*Hexan*), C_6H_{14}.

Of the four known bodies of this composition, two appear to be contained in the Pennsylvania Petroleum.

a. *Normal Hexan.*

Is a mobile liquid, with a boiling point of 69°.5 C., and sp. gr. at 17° C. of 0.663.

This has been obtained in a state of purity, by Pelouze and Cahours (*loc. cit.*), by Warren (*loc. cit.*), and by Schorlemmer (*loc. cit.*).

It was also prepared, though not investigated by Caventou (Compt. rend. LIX, 449).

b. *Aethyl-isobutyl.*

Has a boiling point of 62° C., and sp. gr. at 0° of .7011. This has been obtained in a state of purity, by Warren (*loc. cit.*), who fixes its boiling point at 61°.3 C.

178. *Heptyl Hydrid* (*Heptan*), C_7H_{16}.

Of the four known modifications, two are certainly contained in the Pennsylvania Petroleum, while Schorlemmer considers (Chem. News, Vol. 30, No. 787), that there may be a third.

a. *Normal Heptan.*

Has a boiling point of 98° C., and sp. gr. at 16° C. of 0.7122. This has been obtained in a state of purity by Warren (*loc. cit.*), and Schorlemmer (*loc. cit.*).

b. *Aethyl-amyl.*

Has a boiling point of 90°.4 C.

This has been obtained pure by Warren (*loc. cit*), and by Schorlemmer (*loc. cit.*).

Pelouze and Cahours (*loc. cit.*) found a Heptan of boiling point 92°–94° C., which was certainly a mixture of the two mentioned above.

179. *Octyl Hydrid* (*Octan*), C_8H_{18}.

Of this body, two modifications are found to exist in the Pennsylvania Petroleum.

a. *Normal Octan.*

Boils at 123°–125° C. and has a sp. gr. of 0.7032 at 17° C. Has been obtained from Petroleum, by Warren (*loc. cit.*), who makes the boiling point, however, 127°.6 C.

b. *An isomeric Octan.*

Has been obtained by Warren (*loc. cit.*), who gives 119°, and by Pelouze and Cahours (*loc. cit.*), who give 116° to 118° C. as its boiling point.

180. *Nonyl Hydrid (Nonan)*, C_9H_{20}.

Of this body one modification, which has a boiling point of 150°.8 C., has been obtained by Warren (*loc. cit.*), from the Pennsylvania petroleum, and possibly another by Pelouze and Cahours (*loc. cit.*), with a boiling point of 136°–138° C.

181. *Decatyl Hydrid*, $C_{10}H_{22}$.

182. *Undecyl Hydrid*, $C_{11}H_{24}$

These hydrocarbons are given by Pelouze and Cahours (*loc. cit.*), as occurring in the Pennsylvania petroleum, and they assign to them the boiling points 160°–162° C. and 180°–184° C. respectively. Warren, however, working with much superior methods, concludes that these latter two belong to the next or *Ethylene Series.*

183. *Solid Paraffines.*

That part of the petroleum boiling at a temperature over 300° C. solidifies on cooling to a solid mass, which, when purified, is white and transparent. This substance is called Paraffine, and is proved by the study of its oxydation products (Gill and Meusel, Jour. Chem. Soc., 1868, VI, 466), and by its behavior when heated under pressure (Thorpe and Young, Chem. News, 1873, No. 660, 35) to be a mixture of the higher members of the Marsh-gas Series. They have not as yet been separated. The fusing point of commercial paraffine varies between 45° and 65°.

B. *Ethylene Series (Olefines).*

General Formula, C_nH_{2n}.

The lower members of this series, which constitute so im portant a part of our illuminating gas, do not appear to exist native in the Pennsylvania petroleum, although readily formed from it by distilling it at a red heat.

The only members of this series, isolated from the Pennsylvania petroleum by fractional distillation, are the following:

184. *Decylene*, $C_{10}H_{20}$	Boiling point		174°.9 C.
185. *Undecylene*, $C_{11}H_{22}$	"	"	195°.8 C.
186. *Bidecylene*, $C_{12}H_{24}$	"	"	216°.2 C.

These were all obtained by Warren (*loc. cit.*), and are the same compounds that Pelouze and Cahours (*loc. cit.*) considered as belonging to the Marsh-gas Series.

Warren's claim, however, is only based upon the fact that, using superior methods of distillation, he obtained more constant boiling points, and upon the conformance of his results to a supposed regular difference in boiling point, established by him for this series. The analytical data and the vapor-density determinations would afford too slight a difference to justify us in trusting to them. Until the substitution products of these hydrocarbons—the chlorides, the acetates and finally the derived alcohols—are obtained and studied, we cannot be positive upon this point. Schorlemmer (Ann. der Ch. und Ph. CLXI, 263) has done this already for the lower members of the Marsh-gas Series, and has established definitely the normal paraffines.

No higher series of hydrocarbons appears to have been discovered as yet existing native in the Pennsylvania petroleums; at least no definite results of the kind are before us.

The Benzol Series (C_nH_{2n-6}) which contains members of the greatest importance technically and from which are derived all the group of magnificent dyes known as Anilin Dyes does not appear as yet to be represented. In other Petroleums, however, members of this series have been found. Thus De la Rue and Müller (Chem. Gaz. 1856, 375), found in the Rangoon tar from Burmah, a kind of viscid Petroleum, several members of this series; Bussenius and Eisenstuck (Ann. der Ch. und Ph. CXIII, 151) discovered one member of this series in the Petroleum of Sehnde in Hanover; and lastly Pebal and Freund (Ann. Ch. und Ph. CXV, 19) discovered five members of the series in the Petroleum of Boroslaw in Galicia.

Nor does the Naphthalene Series (C_nH_{2n-12}), from which, as derivatives, we have the beautiful Naphthaline colors, appear as yet to be represented, although De la Rue and Müller (*loc. cit.*) found it to be present in the Rangoon tar.

There is still less probability, of course, that the Anthracene Series (C_nH_{2n-18}) or any member of it would occur native in petroleum. In the products of the destructive distillation of petroleum, however, which we may compare to the coal-tar obtained in the distillation of bituminous coal in the manufacture of illuminating gas, we might readily expect anthracene or some similar compound.

Such a compound has been found in the last products of the distillation of petroleum for paraffine oils, which distillation is pushed finally to a red heat. This appeared in its impure state to be a yellowish gamboge-like mass, which was purified by crystallizing from solution in benzole. It was studied by Prof. Henry Morton, of Steven's Institute of Technology, who (Am. Chem., Nov., 1872) gave it the name *Viridin*, which he afterwards changed to *Thallene*. He investigated specially its fluorescent character. Prof. Geo. F. Barker, of the University of Pennsylvania, informs me that his analysis of this substance shows it to be an isomer of anthracene.

All accounts of its formation concur, however, in saying that it is only formed by distilling petroleum at a red heat, or by what is called destructive distillation, where new compounds are formed by the breaking up of the previously existing ones.

Such is a short sketch of what is known as to the chemical composition of the mixture, occurring so abundantly in the western part of Pennsylvania, and known as Petroleum. There are many points in the results so far obtained that are contradictory and uncertain, and still a larger number of questions with regard to its constitution have not been touched upon. A thorough re-examination of the Pennsylvania Petroleum from the stand-point of Modern Organic Chemistry, and with application of its best methods seems very desirable.

No less desirable on more general grounds would be a thorough technical examination of samples of Petroleum from all the principal districts of the Western counties and a classification of them according to the relative proportions of naphtha, light oils, heavy oils, paraffin, &c., yielded.

Accompanying such examinations should be a complete classification of them according to locality and general neighborhood, noting carefully in such a list the appearance and physical characteristics of each sample, especially the specific gravity, the color, fluidity and consistence in all cases.

The gases escaping from these Petroleum wells should also be collected as carefully as possible, and analyzed by the methods of gas analysis in order to detect, if possible, any gaseous constituents that have escaped previous investigation.

APPENDIX TO HYDROCARBONS.

187. Mineral Coal.

The different kinds of coal, although of infinitely greater importance in the arts than the simple Hydrocarbons, are not entitled from a mineralogical standpoint to rank independently, but must be considered in connection with these latter. The varieties of coal known to us are by no means sharply characterized mineral varieties, either as to uniformity, chemical composition or physical properties.

The origin of coal is a most interesting and important question, but one which is best discussed from a geological point of view, and cannot be taken up here. Reference must be had therefore to some established geological authority. A most excellent resumé of the views on this subject may be found in MacFarlane's invaluable work on the coal regions of America (D. Appleton & Co., 1873, p. 597).

We shall speak of the occurrence of coal in this State under the several varieties of Anthracite, Semi-Bituminous, Bituminous and Cannel Coal, and lastly, Lignite.

a. *Anthracite.*

This variety has a specific gravity of 1.32 to 1.7. Its structure is massive and compact, with a conchoidal fracture in the harder varieties. Its color is jet black and in the hardest kinds even metallic black. It burns with great heating effect and never melts, swells, or forms coke. Its volatile matter, after drying, amounts to only three to six per cent.

With the exception of a few small formations in Massachusetts and Rhode Island, and a certain area in Arkansas and some metamorphic anthracite near Santa Fé, and a small area at the Peak Mountain in Southern Virginia, all the anthracite of this country is found in Pennsylvania.

"For the sake of convenience these anthracite regions are now usually divided into three coal-fields or districts. The first, Southern or Schuylkill coal field is one continuous, unbroken basin, except the semi-detached field of Mine Hill. The second district includes the Shamokin, the Mahanoy and the Lehigh basins, embracing all the other small fields and basins. The third is the large unbroken, northern, or Lackawanna and Wyoming region" (MacFarlane, Coal Regions of Am., p. 12).

The total area of all these anthracite coal basins is as follows:

	Sq. miles.
1. Southern or Schuylkill Basin and Mine Hill	146
2. Shamokin, 50; Mahanoy, 31; and Lehigh Basins, 37	128
3. Wyoming and Lackawanna Basin	198
Total area of all the Anthracite Basins	472

(MacFarlane, p. 15.)

Small quantities of anthracite and semi-anthracite have been found in the Hudson River slates along the foot of the North Mountain, between the Susquehanna and the Potomac; and in the Triassic formation, near Phœnixville, and in York county.

In this connection it may be interesting to record an observation of Prof. Wm. Th. Roepper, of Bethlehem. He found occurring in quartz from Ironton, Lehigh county, peculiar crystalline particles of which one looked like a tetrahedron and others were apparently rectangular plates. These were deeply striated and had a strong metallic lustre, somewhat duller, however, on a surface of fracture. The fracture itself was uneven. These crystals deflagrated with nitre and left a slight ash. He considers them as most probably anthracite.

b. *Semi-Bituminous Coal.*

This name is applied to that bituminous coal which cokes and yields combustible gases, but contains only from 15 to 18 per cent. of volatile combustible matter.

The different sections of the State where this coal is obtained are: the Blossburg region, the Towanda region, the McIntyre region, the Broadtop region, the Frostburg region, and along the summit of the Alleghany Mountain.

c. *Bituminous Coal.*

Under this name are included all coals containing 20 per cent. and upwards of volatile matter, and having rather less than 70 per cent. of carbon, in all the Northern and Western counties of the State.

d. *Cannel Coal.*

This term is used for those varieties of bituminous coal which are of a finer texture and more compact structure than the ordinary kinds. They are especially rich in volatile matter and should therefore be pre-eminently suited for the manufacture of illuminating-gas and for the distillation of oils. In this last respect we can compare them with the bituminous shales used so extensively for the distillation of oils. Indeed their similarity of composition shows them to have had a like origin. In both cases we have rich carbonaceous material deposited in a fine form with different admixtures of earthy matter. Cannel is not extensively used for the manufacture of illuminating-gas because it does not form a coke, and because it contains too much ash.

It is chiefly found replacing a portion of the third or fourth bed from the bottom of the series, Bed C, the Kittanning Coal Bed, or the bed next above it. The principal mine is on Red Bank Creek.

e. *Lignite.*

This is an imperfectly formed coal which still retains the woody structure distinctly, and much more of the composition of wood than the true coals. It is of later geological origin and undoubtedly was formed under circumstances very different from beds of true coal.

Two localities are recorded of the occurrence of Lignite in this State. The first is described by Prof. J. P. Lesley, (Proc. Am. Phil. Soc. Vol. IX, 72) as occurring at the Pond Iron Ore Bank, near Chambersburg in Franklin county.

The other is described by Dr. Leidy (Proc. Acad. Nat. Sc. Phil., 1861, 77) as found on Plymouth Creek, near Norristown.

INDEX

To the Mineral Species, Localities, and relationships, and to the names of Chemists and Mineralogists mentioned in Dr. Genth's Report of 1874.

LABORATORY OF THE UNIVERSITY OF PENNSYLVANIA,
December 31*st*, 1875.

Professor J. P. LESLEY,

State Geologist:

SIR:—I beg to submit in the following pages my Report on the work done during the year, a report of progress. From my last Report it was evident that some doubt existed in regard to the constitution and form of many of our mineral species. By numerous analyses, the chemical composition of many has since been established, but nothing has been done yet to study their crystallographic peculiarities.

I have also commenced the examination of the mineral waters of the State, but regret that some very important springs have not yet been analyzed. This is owing, in part, to the very elaborate investigation required to settle the true character of several of reputation, which consumed so much time, that none was left for others, perhaps more deserving. It is my hope that I shall be able to do justice to these in the coming year.

Particular attention was paid to the study of the rock masses occurring in the State. Their investigation is attended with many difficulties and requires much time. Henceforth I intend to devote the greater part of it to this very important subject, in order to be able to embrace the results of the labors of some of your assistants and my own, in a Lithology of Pennsylvania.

I give in my Report the results of the analyses of several of the natural gases of the State, which have assumed such importance in the manufacture of iron. These analyses have been made with great care by Prof. Samuel P. Sadtler. I hope that other gases which have not yet been analyzed, may be fully investigated.

Nothing has yet been done with petroleum, the great staple of Pennsylvania. With the exception of a few statements compiled by Dr. Sadtler, from the investigations of a few varieties by other chemists, and given as an appendix to my Report of 1874, nothing is known. An investigation of typical specimens of the various varieties from different geological horizons, may throw a great deal of light upon this subject, both from a geological and a chemical point of view. I believe that it is very desirable that this very important work should be commenced at an early day.

I have the honor to remain,

Your obedient servant,

F. A. GENTH.

CHAPTER I.

NATIVE ELEMENTS.

Gold. [1. *page* 2 *B.*]

Major Samuel L. Young, of Reading, has a fragment of quartz, containing a small particle of gold of about the size of a No 5 shot, crushed, which is said to have been found in the neighborhood of Reading, by the late Jno. P. Miller.

Dr. W. J. Hoffman, of Reading, who kindly furnished me with this information, states that the matrix appears to correspond with the rocks found north-east of the city, but that he has never felt perfectly satisfied regarding the specimen in question.

Although somewhat doubtful, I give this additional locality, as it may lead to the discovery of larger and perhaps workable quantities of this metal.

Iron. [4. *page* 6 *B.*]

The small portion of the meteoric iron from Pittsburg, referred to in my report B, of 1874, has been preserved in the Yale College cabinet. Mr. Edward S. Dana, the curator of this collection, presented me with a small fragment, which I submitted to a fuller investigation.

Its specific gravity, which Shepard gave as 7.380, was found to be 7.741, the average of three closely corresponding determinations by Dr. Koenig, Dr. Headden, and myself After polishing and etching with dilute nitric acid, it presents Widmannstätten's figures, which are produced by inclosed schreibersite. In the section which has been made, it happens that most of the exceedingly minute schreibersite crystals are cut across, and are seen as small dots on a frosted surface; some appear as minute needles, arranged in parallel lines, like the trees in an orchard. A few elongated patches of a whiter iron-nickel alloy are also visible.

The analysis of a somewhat oxydized piece, gave the following composition:

Iron	=	92.809
Copper	=	0.034
Cobalt	=	0.395
Nickel	=	4.665
Manganese	=	0.141
Sulphur	=	0.037
Phosphorus	=	0.251
		98.332

0.251 per cent of phosphorus are equal to about 1.8 per cent of schreibersite.

CHAPTER II.

COMPOUNDS.

SULPHIDES.

Pyrite. [19. *page* 19 *B.*]

In sinking a shaft for coal, four miles west of Parker, an interesting variety of pyrite was found in the fire-clay. It consisted of a nodular mass of this mineral, which on being broken was found to be a geode, lined with minute octahedral crystals of pyrite, coated with a film of galenite, and also containing some sphalerite or zincblende.

The specimen was kindly communicated by Mr. D. Jones Lucas, resident engineer for the Empire Transportation Company at Parker.

Globular pyrite, radiating from the centre, in balls of from half an inch to several inches in diameter, occurs in the limonite beds on the farm of B. Boganstoss, in Centre township, Franklin county. I am indebted to Mr. H. W. Hollenbush, of Reading, for this information and for specimens.

Arsenopyrite. [24. *page* 23 *B.*]

The occurrence of this mineral at the Perkiomen mines, which I considered doubtful, is confirmed by a statement from Dr. W. J. Hoffman, of Reading, who informed me that, while at school at Collegeville, Montgomery county, the students occasionally visited the copper mines near Shannonville, and found various quantities of mispickel, which, at that time, was not uncommon.

CHAPTER IV.

FLUORINE COMPOUNDS

Fluorite. [30. *page* 29 *B.*]

Very interesting specimens of fluorite have lately been found by Mr. Theo. D. Rand, at Frankford. It occurs in small octahedral crystals of a deep purple color, frequently inclosed in, and forming a nucleus of colorless fluorite. The crystals are sometimes cavernous, and are often surrounded by a white earthy substance which, on examination, was proved to be *amorphous* silicic acid. When first received it looked very much like the silicic acid which separates on the decomposition of fluoride of silicon by water, which appearance suggests a similar origin.

Dr. W. J. Hoffman communicates, as a new locality, De Turk's farm in Exeter township, four miles east of Reading, where it occurs sparingly in coarse granular, occasionally crystallized masses of a violet and purple color, associated with calcite.

Mr. H. W. Hollenbush has found it in imperfect cubical crystals and crystalline granular masses of a pale grayish white or pink color, associated with lamellar barite at Samuel Plum's farm, Franklin county.

Violet fluorite has been observed by Prof. Persifor Frazer, Jr., as a coating of the upper surface of auroral limestone in Thomas' quarry, Pine Grove, Cumberland county.

CHAPTER V.

OXYGEN COMPOUNDS.

Water. [32. *page* 31 *B.*]

The analysis of the mineral waters of the State was commenced during last summer, and of some of the more important springs very elaborate examinations were made; in *all cases, of water which was undoubtedly genuine, and collected by gentlemen in the employ of the Geological Survey.*

A number of important springs have as yet not been visited, and their waters will be analyzed as soon as it can be done without interrupting other more important investigations.

Prof. Persifor Frazer, Jr., collected and sent for analysis *five* gallons of the so-called "Lithia" spring, on Mr. Stremmel's property, about half a mile north of the "*Gettysburg Katalysine*" water, of which latter I gave an analysis in my Report B for 1874, page 28.

a. *Stremmel's Gettysburg Lithia Spring.*

It contains in one gallon of 231 cubic inches:

Sulphate of magnesia	=	3.29559	grains.
" lime	=	0.48243	"
" potash	=	0.15399	"
Chloride of sodium	=	0.31836	"
" lithium	=	trace	
Bicarbonate of iron	=	0.04203	"
" manganese	=	0.00485	"
" magnesia	=	5.82961	"
" lime	=	9.95838	"
" soda	=	3.37602	"
Phosphate of lime	=	0.00963	"
Alumina	=	0.02425	"
Silicic acid	=	1.75473	"
		25.24987	"

Five litres of this water were examined for fluorine, boracic acid, baryta and strontia, but neither could be detected.

The mineral waters from Huntingdon county, near Saltillo, East Broad Top railroad, seventeen miles from Mt. Union, on the Pennsylvania railroad, have been collected and forwarded by Mr. Charles A. Ashburner, viz:—

b. *S. M' Vitty's Spring* is located in the Clinton upper olive shales, north-west of Jack's mountain anticlinal.

c. *C. R. M'Carthy's Spring*, at the lower horizon of the Lewistown limestone, north-west of Jack's mountain anticlinal.

They contain in one gallon of 231 cubic inches:

		M'Vitty's spring.		M'Carthy's spring.	
Sulphate of magnesia	=	0.00456	—	41.79795	grains.
" lime	=	———	—	72.19660	"
" soda	=	1.00664	—	7.79412	"
" potash	=	0.15624	—	0.22291	"
Chloride of sodium	=	0.06329	—	0.20571	"
Bicarbonate of iron	=	0.14022	—	0.08108	"
" magnesia	=	1.87476	—	0.88262	"
" lime	=	9.84013	—	22.24300	"
Phosphate of lime	=	trace	—	trace	
Silicic acid	=	0.59007	—	1.16846	"
Hydrosulphuric acid	=	0.01490	—	0.01589	"
		13.69081	—	146.60834	"

d. The water from the "*Fayette Spring*," situated in Wharton township, Fayette county, near the eastern slope of the "Chestnut Ridge," (usually but incorrectly called "Laurel Ridge,") nine miles east by south of Uniontown, has been collected and forwarded by Mr. Charles A. Young.

It contains in one gallon of 231 cubic inches:

Sulphate of magnesia	=	0.25472	grains.
" lime	=	0.05542	"
" soda	=	0.19965	"
" potash	=	0.11525	"
Chloride of sodium	=	0.08522	"
Bicarbonate of iron	=	1.06709	"
" manganese	=	0.04795	"
" magnesia	=	1.53414	"
" lime	=	9.33441	"
Phosphate of lime	=	0.04822	"
Alumina	=	trace	
Silicic acid	=	1.19690	"
Nitrous acid	=	trace	
Carbonic acid (free)	=	0.38284	"
		14.32181	"

The mineral waters at *Cresson*, Cambria county, were collected by Mr. Franklin Platt, who states that the (e) *Iron Spring* comes off the bench of the Upper Freeport Coal bed; the (f) *Alum Spring* has apparently the same geological position and is opened at the mouth of an old coal drift, not 50 yards from the Iron Spring. The (g) *Magnesia Spring* is higher up in the measures.

One gallon of 231 cubic inches contains:

		e. *Iron Spring.*		f. *Alum Spring.*	
Sulphate of ferric oxide	=	trace	—	33.38970	grains.
" alumina	=	1.60466	—	21.20498	"
" ferrous oxide	=	23.47923	—	16.25273	"
" magnesia	=	22.58007	—	27.69855	"
" lime	=	48.91824	—	40.20179	"
" lithia	=	trace	—	0.04693	"
" soda	=	1.64331	—	0.70398	"
" potash	=	0.32405	—	0.42622	"
Chloride of sodium	=	0.04063	—	0.02336	"
Bicarbonate of iron	=	5.03471	—	3.74756	"
" manganese	=	———	—	trace	
" lime	=	3.52946	—	———	"
Phosphate of lime	=	0.02914	—	trace	
Silicic acid	=	1.20832	—	1.86794	"
		108.39182	—	145.56374	"

		g. *Magnesia Spring.*	
Sulphate of lime	=	0.10912	grains.
Chloride of magnesium	=	0.55962	"
" calcium	=	1.30444	"
" sodium	=	1.22974	"
Bicarbonate of iron	=	0.01753	"
" manganese	=	trace	
" magnesia	=	0.41434	"
" lime	=	0.02252	"
" soda	=	1.42582	"
" potash	=	0.20671	"
Phosphate of lime	=	0.00408	"
Alumina	=	0.00876	"
Silicic acid	=	0.91455	"
Nitrous acid	=	trace	
Carbonic acid (free)	=	0.66390	"
		6.88113	

h. *Minnequa Spring, Bradford County.*

The water for analysis was collected and forwarded by Mr. Franklin Platt, on the 23d of September, in a carboy holding about 12 gallons. As the results of the analysis of this *unquestionably genuine* water were at variance with the results of another analyst, I thought it advisable to make another analysis. Mr. W. G. Platt, therefore proceeded to Minnequa, and collected another carboy of the same capacity, which was received on the 12th of December. Both analyses agreed very closely. It was evident, however, that the latter water was a trifle weaker, giving 7.57162 grains of solid constituents, instead of 7.78485 as in the former; 0.02753 grains of manganous oxide, instead of 0.02873, and 0.04323 boracic acid instead of 0.05521, which latter result was obtained in two trials with the first water. Although these discrepancies existed, probably on account of a slight contamination with rain water, the two waters were essentially the same; the average results of my analyses are therefore given as follows:

Solid constituents	=	7.67824 grains per gall.
Sulphate of baryta	=	0.00175 grains.
Chloride of lithium	=	faint trace
" sodium	=	0.19209 grains.
Bicarbonate of zinc	=	0.01157 "
" manganese	=	0.06299 "
" iron	=	0.04204 "
" magnesia	=	1.58988 "
" lime	=	6.52477 "
" baryta	=	0.01380 "
" soda	=	1.33046 "
" potash	=	0.13885 "
Borate of magnesia	=	0.07980 "
Phosphate of lime	=	0.01231 "
Nitrite } Nitrate } of ammonia	=	0.00025 "
Alumina	=	0.00253 "
Silicic acid	=	0.74194 "
Hydrosulphuric acid	=	0.01390 "
		10.75893 "

Hematite. [35. *page* 33 *B.*]

Mr. G. W. Hathaway sent me under date of November 19th, 1875, a specimen of the "Tioga ore," with some of the metal reduced from it, together with a number of articles from the *Tioga County Express*, with reference to it. As these communications advocate the opinion that the Tioga ore contains "something" peculiar, new or unknown to chemists, which "something" would produce a very superior quality of iron,* I have made a careful analysis of the sample which Mr. Hathaway has sent; I have also tested the metal produced from it.

The Tioga Ore contains according to my analysis:

Silicic acid	=	27.69
Phosphoric acid	=	0.93
Ferric oxide	=	55.18
Alumina	=	9.39
Magnesia	=	0.93
Lime	=	1.72
Alcalies, &c., (by difference)	=	0.73
Water	=	3.43
		100.00

The ore contains, therefore, 38.63 per cent of iron and 0.41 per cent of phosphorus.

All that can be said about the Tioga ore is that it is an inferior quality of iron ore, without any strange or valuable metal other than iron.

Göthite. [49. *page* 48 *B.*]

Prof. Persifor Frazer, Jr., observed large quantities of this mineral with hematite in copper colored unctuous scales in the Altland Mine.

Limonite. [50. *page* 49 *B.*]

Through a printer's error in the analysis of Mr. Sidney Castle (4) of the limonite from Middletown, Delaware county, published in my Report for 1874, page 53 — 6.90 per cent are extended in the line of quartz instead of that of silicic acid, which should be corrected.

* Since the above was written, Mr. Hathaway sent me other newspaper articles, in which the astonishing announcement was made that "spiritual manifestations" had revealed a way by which the *new metal,* called "motor," could be separated. I should perhaps not notice in any manner such statements, but, seeing that great efforts are made to play upon the credulity of the public, I will merely say that the whole matter has no sound basis in fact, and no such statements ought to change my opinion respecting the Tioga ore.

This variety has been analyzed by Mr. W. H. Melville of the Senior Class in Harvard College, Cambridge, Mass., and described under the name "melanosiderite," by Prof. Josiah P. Cooke, Jr., (Proc. Am. Acad. Arts and Sciences, 1875, 451,) as a *new* mineral species. The improbability of Prof. Cooke's suggestion induced me to submit to a thorough investigation specimens which I had myself collected, as well as others received from Mr. W. W. Jefferis, and by him pronounced as *identical* with those sent to Prof. Cooke.

The specific gravity in different pieces was found to be as follows:

3.140—3.165—3.186—3.203 and 3.326.

The water in several pieces taken from the mine for over one year was found as follows:

15.95—17.19—17.70 per cent.

The specimen, which gave 15.95 per cent, contained, when dried over sulphuric acid,

During	24 hours	11.48 per cent of water.
"	2 days	10.64 "
"	2 weeks	10.58 "
Over	2 months	10.12 "

After drying for two weeks it lost, when exposed for 6 hours to a temperature of 100°C—2.16 per cent, and on ignition gave in addition 8.42 per cent of water.

The specimen, which "air-dry" gave 17.70 per cent contained, after drying over sulphuric acid during one month, 10.45 per cent.

A quantity of the mineral was allowed to remain in dilute chlorhydric acid (1: 10) during two months. After this time it was found that it was decomposed into yellowish—brown and yellowish—white amorphous silicic acid, a semi-transparent brownish and black scaly, apparently crystalline mineral, perhaps göthite, and a solution, containing 0.73 per cent of silicic acid, together with the greater part of the basic constituents.

From these experiments no doubt can remain that "melanosiderite" is *not a good species*, but an *impure* variety of ferric hydrate, probably limonite, under which head I considered it before. The analyses of two different specimens gave:*

	a. Dried over Sulphuric acid for 24 hours (air-dry=15.95.)		b. Calculated for —dried 2 months.		c. Air-dry.		d. Calculated for —dried 1 month.
Ignition	= 11.48	—	10.42	—	17.19	—	10.45
Silicic acid	= 7.38	—	7.47	—	7.27	—	7.79
Phosphoric acid	= 0.26	—	0.26	—	trace	—	trace
Alumina	= } 79.53	—	80.37	—	2.71	—	2.91
Ferric oxide	= }				73.00	—	78.25
Cobaltic oxide	= 0.39	—	0.39	—	0.16	—	0.17
Manganic oxide	= 1.08	—	1.08	—	0.40	—	0.43
	100.12		100.00		100.73		100.00

Psilomelane. [53. *page* 53 *B.*]

Stalactitic, botryoidal and reniform masses have been found in large quantities at Ironton, Lehigh County.

The importance of the discovery of a rich manganese ore made an analysis desirable. A specimen received from Prof. Fred'k Prime, Jr., was analyzed in the Laboratory of the University of Pennsylvania, by Mr. Henry Pemberton, Jr., who found the spec. gr. = 4.095, and the composition as follows:

Manganese binoxide	=	84.88	contains	31.22	oxygen.
Manganous oxide	=	3.77	"	0.85	"
Cobaltous oxide	=	1.68	"	0.36	"
Niccolous oxide	=	trace			
Magnesia	=	0.79	"	0.32	"
Lime	=	1.90	"	0.54	"
Baryta	=	trace			
Soda	=	0.19	"	0.05	"
Potash	=	3.50	"	0.59	"
Water	=	4.38	"	3.89	"
Phosphoric acid	=	trace			
		101.09			

From this analysis it is evident that the Ironton mineral is a mixture of psilomelane and pyrolusite, containing 56.58 per cent of metallic manganese.

* Compare also Castle's analysis—page 51, 4.

Wad. [54. *page* 53 *B.*]

Dr. W. J. Hoffman gives me as a new locality for wad in mammillary concretions the Morgantown Road, ½ mile south of Reading.

Quartz. [56. *page* 55 *B.*]

Quartz, in crystals, with pyramidal terminations occurs, according to Prof. Persifor Frazer, Jr., in the coarse crystalline syenitic rocks near Harman's blacksmith shop, six miles southeast of Dillsburg, York county.

In fine crystals, from one-half to one and one-half inches in length, sometimes doubly terminated, it is found on the Marshall coal tract, three miles west of Swatara, in Schuylkill county; in crystals, from two to four inches long, it occurs at Heckel's forge.

Drusy quartz, occasionally tinted milky, pale blue, pink or ferruginous, is found near Pricetown, in Berks county.

Beautiful masses of botryoidal *chalcedony* occur at the Perkiomen copper mines.

Hornstone, *silicified wood* and *basanite* have been found near Friedensburg, in Oley township, east of Reading, in Berks county. *Jasper* has been found in yellow masses near Pricetown, Berks county.

Arrow-heads, scrapers, spear-heads and flakes of various shades of jasper, are very abundant near the water-courses; at the corner of Twelfth and Spruce streets, Reading, large quantities of flints are found, and there has either been a "workshop," or a camping ground, used over a prolonged period.

I am indebted to Dr. W. J. Hoffman for these additional data about quartz.

Opal. [57. *page* 61 *B.*]

I have already mentioned under fluorite the peculiar amorphous silica, which surrounds the fluorite at Frankford. Its solubility in dilute potash solution proves it to be a variety of *opal.*

Pyroxene. [60. *page* 64 *B.*]

a. I have analyzed the pyroxene variety from Bailey's Quarry, mentioned on page 65 B.

Its specific gravity was found to be=3.229.

It contains:—

Silicic acid	=	52.19
Carbonic acid	=	1.79
Alumina	=	0.81
Ferrous oxide	=	0.99
Magnesia	=	17.36
Lime	=	26.07
Water	=	0.73
		99.94

The purest material, in perfect cleavage masses, was selected for this analysis, but notwithstanding the greatest care it showed an admixture of 3.74 per cent of dolomite;—after deducting this, the constituents give the atomic ratios corresponding with the variety "malacolite."

b. I have also analyzed a beautiful variety of pyroxene, occurring in cleavage masses, often over one inch across and several inches in length, at Godshall's mine, Alsace Township, Berks Co.

It has a dark greenish-black color, and sometimes shows slightly a sub-metallic bronze lustre, especially when it is slightly altered. The results of analysis are as follows:

Specfic gravity, = 3.317.

Silicic acid	=	51.42
Alumina	=	1.94
Ferrous oxide	=	16.41
Manganous oxide	=	0.67
Cobaltous oxide	=	0.06
Cupric oxide	=	0.07
Magnesia	=	9.27
Lime	=	18.97
Ignition	=	1.36
		100.17

c. The brownish vitreous mineral with distinct cleavage in one direction, forming, with labradorite, the chief constituents of some of the Pennsylvania "*trap-rocks*," was picked out as carefully as possible from a specimen of trap from the north side of "Devil's Den," about 3 miles S. of Gettysburg, collected by Prof. P. Frazer, Jr. The analysis gave me the following results:

Silicic acid	=	51.64
Alumina	=	4.23
Ferrous oxide	=	16.04
Magnesia	=	15.93
Lime	=	10.05
Soda	=	0.46
Potash	=	0.16
Ignition	=	0.72
		99.23

It appears from these results that this pyroxene was slightly contaminated with labradorite, the analysis of which, from the same rock, will be given below.

Garnet. [65. *page* 72 *B.*]

A brown garnet occurs in the iron ores at Dillsburg, York Co., and at W. Hartman's, near Reading.

Epidote. [68. *page* 78 *B.*]

Crystals of epidote an inch or more in length, with the usual planes, have been picked up in large numbers in the gullies leading from Cemetery Hill, near Gettysburg, as Prof. P. Frazer, Jr., informs me.

It is also found at Haines' mine, near Pricetown, and associated with garnet at W. Hartman's, near Reading, in Berks Co.

Feldspar Group. [*page* 87 *B.*]

I have commenced the analyses of numerous doubtful feldspars. The following have been completed:

Labradorite. [76. *page* 88 *B.*]

The feldspathic constituent of some specimens of "trap-rocks" from the neighborhood of Gettysburg, collected by Prof. Persifor Frazer, Jr., has been analyzed by me with results which leave no doubt that feldspar is *labradorite.*

The specimens came (a) from the N. side of Devil's Den, and and (b) from the W. side of Round Top—both about 3 miles S. of Gettysburg.

I found:

		a.	b.
Silicic acid	=	54.05	53.85
Alumina	=	28.81	28.91
Ferrous oxide	=	1.36	1.16
Magnesia	=	0.26	0.22
Lime	=	11.05	11.79
Soda	=	3.36	3.23
Potash	=	0.59	0.77
Ignition	=	0.45	0.76
		99.93	100.69

The presence of the small percentage of ferrous oxide and magnesia is accounted for by a slight contamination with pyroxene.

It being intended to make a full investigation of *all* the rocks of the State in a separate " Lithology of Pennsylvania," it may be premature to communicate results only partly com pleted; but as this report is intended to enumerate the amount of work done during the year, I will insert a few analyses of "trap-rocks" which I have made, and which are of particular interest in connection with the investigations recently made by Prof. J. D. Dana and Mr. Geo. W. Hawes, of Yale College, of similar rocks from other States.

a. Specimens from a small trap-dyke in the serpentine near Moro Phillips' chrome mine, West Nottingham township, Chester county, appear to consist, when examined in thin sections, principally of triclinic feldspar, pyroxene, a small quantity of chrysolite and magnetite; the rock is probably "basalt."

The spec. gr. is=2.989. When treated with dilute chlorhydric acid, 40.30 per cent are dissolved.

Analyses were made: a, of the rock itself; b, of its soluble portion; and c, of the insoluble portion.

		a.		b.		c.
Silicic acid	=	51.25	—	43.16	—	55.59
Titanic acid	=	0.17	—	0.43	—	——
Phosphoric acid	=	0.04	—	0.10	—	——
Alumina	=	14.84	—	21.16	—	9.92
Chromic oxide	=	0.14	—	——	—	0.23
Ferric oxide	=	4.21	—	2.98	—	5.03
Ferrous oxide	=	6.46	—	8.40	—	4.60
Manganous oxide	=	0.11	—	0.28	—	trace
Magnesia	=	9.32	—	6.90	—	10.40
Lime	=	10.14	—	10.20	—	11.11
Soda	=	1.88	—	2.68	—	2.31
Potash	=	0.61	—	0.33	—	0.67
Water	=	1.35	—	3.38	—	——
		100.45		100.00		99.86

b. Fine-grained trap—probably *dolerite*—from Beeler's farm, 2 miles S. W. of York, has been collected and submitted for analysis by Prof. Persifor Frazer, Jr. I obtained the following results:

Specific gravity = 3.007.

Silicic acid	=	52.53
Titanic acid	=	0.32
Phosphoric acid	=	0.15
Alumina	=	14.35
Ferric oxide	=	5.93
Manganous and cupric oxides	=	traces
Ferrous oxide	=	5.45
Magnesia	=	7.99
Lime	=	10.27
Soda (faint trace of Lithia)	=	1.87
Potash	=	0.92
Sulphur	=	0.03
Ignition	=	1.23
		101.04

c. I have collected and analyzed two varieties of trap, probably *dolerite*, adjoining the magnetic iron ore at the Cornwall Ore Bank. The following are the results:

		a. Coarse-grained.		b. Fine-grained.
Specific gravity	=	3.009	—	2.999
Silicic acid	=	53.09	—	53.88
Titanic acid	=	0.90	—	1.09
Phosphoric acid	=	0.16	—	0.10
Chromic oxide	=	trace	—	0.06
Alumina	=	14.97	—	14.33
Ferric oxide	=	6.13	—	3.23
Ferrous oxide	=	4.47	—	6.54
Manganous oxide	=	trace	—	0.09
Magnesia	=	6.50	—	7.25
Lime	=	11.24	—	10.92
Soda	=	2.10	—	2.08
Potash	=	0.69	—	0.96
Ignition	=	0.40	—	0.23
		100.65		100.76

I have commenced the analyses of the rocks, which form the trap-dykes, entering Pennsylvania from New Jersey near New Hope, Bucks County, and those crossing to N. Pa. R. R. near Quakertown, and those from the neighborhood of Gettysburg.

I hope to give a full investigation of these and others in a "Lithology of Pennsylvania," for which I am at present collecting material.

Oligoclase. [77. *page* 89 *B.*]

Several feldspars, which belong to this species, have been analyzed by me:

1. A pale greenish-white feldspar, with triclinic striation, which occurs, associated with a reddish orthoclase, in the granite veins of the gneissic rocks at Frankford.

2. An almost colorless vitreous variety, with deep triclinic striæ, from the jefferisite locality at West-town near West-Chester. It resembles closely the variety from Dismal Run, analyzed by Mr. Haines (Report B, page 90).

3. A feldspar of white color, showing on the cleavage planes a very delicate striation, occurs, together with cassinite, at the Blue Hill, Delaware County. It is, like the latter, surrounded with granular quartz, but shows the composition of oligoclase.

4. To oligoclase probably belongs a pale bluish, dull feldspathic mineral, which occurs near the soapstone quarries, with chlorite, talc, garnet and menaccanite.

As will be seen from the oxygen ratios given below, which, for a pure oligoclase, should be 1: 3: 9, that the analyses show some deviation from it, propably, in part, owing to mechanical admixtures of foreign substances, in part perhaps to an interlamination with albite.

Where the discrepancies appeared to be too great, the results of the analysis were verified by a second.

The following are the results:

	1 Frankford.	2 Westtown. a.	2 Westtown. b.	3 Blue Hill.	Soapstone Quarry. a.	Soapstone Quarry. b.
Specific gravity =	2.712	2.689		2.658	2.670	
Silicic acid =	63.63	61.04	61.65	64.51	61.82	62.55
Alumina =	22.79	24.25	24.13 (Alumina and Ferric oxide)	21.54	25.52	25.43 (Alumina and Ferric oxide)
Ferric oxide =	——	0.07		0.05	0.47	
Ferrous oxide =	0.43	——	——	——	——	——
Manganous oxide =	——	——	——	——	0.04	not det.
Magnesia =	0.09	0.29	0.28	——	0.22	not det.
Lime =	5.51	5.05	5.25	3.31	1.81	1.87
Baryta =	——	——	——	0.10	——	——
Soda =	6.87	7.92	not det.	9.37	8.03	not det.
Potash =	0.75	0.58	not det.	1.25	2.40	not det.
Ignition =	0.42	0.54	not det.	0.32	1.43	not det.
	100.49	99.74		100.45	101.74	

The oxygen ratios of RO : R_2O_3 : SiO_2 are in:

		RO		R_2O_3		SiO_2
1	=	1	:	2.93	:	9.35
2	=	1	:	3.2	:	10.17
3	=	1	:	2.79	:	9.54
4	=	1	:	3.9	:	10.5

Orthoclase. [79. *page* 92 *B.*]

From observations already made by Mr. Th. D. Rand, and confirmed by me, it became evident that Dr. Isaac Lea's varieties, the "lennilite" and "delawarite,"—mentioned in Report B, page 93,—were identical. I have analyzed a very carefully selected sample, showing a greenish color and vitreous lustre on one plane and a grayish-white color and strong, satin-like, pearly lustre upon the other. The results of the analysis give the composition of orthoclase—containing the alkalies potash and soda in the atomic ratio 1: 0.8.

Specific gravity, = 2.619.

Silicic acid	=	65.77 per cent.
Alumina	=	19.21
Ferric oxide	=	0.26
Manganous oxide	=	trace
Magnesia	=	trace
Lime	=	0.16
Baryta	=	0.57
Soda	=	4.88
Potash	=	9.11
Ignition	=	0.20
		100.16

Several specimens of Dr. I. Lea's variety, "cassinite," from Blue Hill, Delaware county, were analyzed by me and gave highly interesting results, showing it to contain a considerable quantity of baryta. The following are the results:

Specific gravity = 2.692

		1.		2.		3.
Silicic acid	=	62.88	—	62.68	—	62.25
Alumina	=	20.10	—	20.04	—	19.76
Ferric oxide	=	0.08	—	0.08	—	0.21
Magnesia	=	not deter'd	—	0.02	—	trace
Lime	=	0.16	—	0.15	—	0.27
Strontia	=	trace	—	trace	—	trace
Baryta	=	3.79	—	3.75	—	3.60
Soda	=	4.06	—	4.06	—	4.80
Potash	=	8.83	—	8.83	—	9.18
Ignition	=	0.19	—	0.17	—	0.22
		100.09		99.78		100.29

The atomic ratio of RO: R_2O_3 : SiO_2, in the average of the two first analyses is=1 : 3.13: 11.20; in the third analysis =1: 2.87: 9.98.

Fuller examinations, especially with reference of the crystallographic characters of the cassinite are necessary to decide the exact position of this mineral.

It has been stated in my report of 1874, B. 94, that orthoclase occurs at Van Arsdale's quarry in Bucks county, in cleavable masses; sometimes opalescent with rich blue colors.

I have analyzed one of the latter and obtained highly interesting results. The material for analysis appeared quite uniform throughout, and was of a dark grey color with blue opalescence. The particles showed *distinct striation.* The analysis gave:

Specific gravity, = 2.497.

Silicic acid	=	64.93
Alumina	=	18.58
Ferric oxide	=	0.49
Magnesia	=	0.12
Lime	=	1.77
Soda	=	3.04
Potash	=	10.44
Ignition	=	1.11
		100.48

The atomic ratio of the constituents is for RO: R_2O_3 : SiO_2 =1: 2.81: 10.84=1.07: 3: 11.58, agreeing nearly with orthoclase, whilst it shows the striation of a triclinic feldspar. By the examination of the crystalline structure, it is yet to be ascertained whether this opalescent variety is not a mixture of orthoclase with either labradorite or oligoclase, as the apparently pure orthoclase without striation and perfectly rectangular cleavage is found at the same locality, or a *new* feldspar, showing the composition of orthoclase with a triclinic form.

Chabazite. [96. *page* 108 *B.*]

Small rhombohedral crystals of chabazite of a yellowish brown color have been found by Mr. Lewis Palmer, in cavities of a hornblendic rock, in the neighborhood of Media, Delaware county.

Stilbite. [97. *page* 109 *B.*]

The white radiated stilbite from Frankford, has been analyzed in the laboratory of the University of Pennsylvania, by Mr. Walter A. Fellows, who found:

Silicic acid	=	56.74 per cent.
Alumina	=	16.44
Lime	=	8.29
Soda	=	0.14
Potash	=	0.06
Water	=	18.90
		100.57

Pyrophyllite. [101. *page* 112 *B.*]

A beautiful specimen of this mineral in radiating crystalline masses of a greenish-white color, has, associated with talc, lately been found at Prince's soapstone quarries, near Lafayette, Montgomery Co.

Hallite. [110. *page* 129 *B.*]

Prof. Josiah P. Cooke, Jr., and F. A. Gooch have published in the Proceedings of the American Academy of Arts and Sciences of 1875, (pages 453 ff,) in an article "*on two new varieties of vermiculites*," the description and analysis of that from Lenni, Delaware Co., mentioned in my Report for 1874 (page 130).

The composition of the mineral, after being dried at 100° C. until the weight was constant, is—as the mean of three analyses—as follows:

Silicic acid	=	38.03 per cent
Alumina	=	12.93
Ferric oxide	=	7.02
Ferrous oxide	=	0.50
Magnesia	=	29.64
Lithia, Potash	=	traces
Water	=	11.68
		99.80

I may here state that the locality is erroneously called "Lerni," instead of "Lenni."

Chloritoid. [117. *page* 135 *B.*]

Prof. Persifor Frazer, Jr., has observed at Whitestown, 4 miles N. W. from Petersburg, Adams county, dark greenish-black scales or small plates of not over $\frac{1}{8}$ inch in size, which occur disseminated through a (?) chloritic schist.

The carefully selected and *almost* pure mineral has a spec. gr.=3.197, and had the following composition:

Silicic acid	=	28.19 per cent
Alumina	=	37.67
Ferric oxide	=	3.12
Ferrous oxide	=	22.21
Manganous oxide	=	trace
Magnesia	=	2.28
Water	=	6.53
		100.00

The results of the analysis were calculated for the pure mineral, after deducting 5.91 per cent of silicic acid, which is probably present as *quartz*, and 1.92 per cent of titanic acid, most likely present as *rutile.*

Chloritic Mineral from Willets.

The analysis of a fine, scaly, somewhat granular, dark green chloritic mineral, which Prof. Persifor Frazer, Jr., has found at Willets, south of Littlestown, Adams County, gave results showing it to be a still more basic compound than thuringite, which it resembles with this difference only that the scales are somewhat larger and their aggregation less tough. It contains, after subtracting 5.26 per cent of quartz, as follows:

Specific gravity, = 3.131.

Silicic acid	=	22.46	contains	11.66		oxygen.
Alumina	=	22.84	"	10.66	=11.52	"
Ferric oxide	=	2.85	"	0.86		
Ferrous oxide	=	35.70	"	7.94		
Manganous oxide	=	0.10	"	0.02	10.40	"
Magnesia	=	6.11	"	2.44		
Water	=	9.94	"	8.84		"

The oxygen ratio nearly corresponds with: $5\ RO+2\ R_2O_3+3\ SiO_2+4\ H\ O$, which might be represented by $3\ (RO,\ SiO_2)+2\ (RO,\ R_2O_3)+4\ H_2O$.

There are several other chloritic minerals in the State which require fuller investigation.

Pyromorphite. [121. *page* 139 *B.*]

Mr. Henry Pemberton, Jr., has analyzed in the laboratory of the University of Pennsylvania, the green granular pyromorphite from the Wheatley mine near Phœnixville, and has obtained the following results:

Specific gravity = 7.117.

Phosphate of lead	=	90.35
Chloride of lead	=	9.11
Chromic oxide	=	0.21
Alumina and ferric oxide	=	0.55
Quartz	=	0.25
		100.47

Vivianite. [124. *page* 141 *B.*]

This mineral has been found in crystalline smalte-blue coatings upon fire-clay, in M'Kean county.

Wavellite. [127. *page* 142 *B.*]

Seams of white wavellite of a radiated structure have been observed by Mr. Charles A. Ashburner, in the Oriskany sandstone at Orbisonia, Huntingdon county.

Barite. [135. *page* 145 *B.*]

White lamellar barite, associated with fluorite, has been found on S. Plum's farm, Franklin Co., by Mr. H. W. Hollenbush.

It is also found in colorless crystals and crystalline masses in limonite in the "chert vein" in the lower Oriskany at Sandy Ridge, near Orbisonia, Huntingdon county

Calcite. [154. *page* 152 *B.*]

Dr. W. J. Hoffman kindly furnished me with the following information about calcite:

Large rhombohedral pink calcite is found one mile north of Reading, within the city limits.

Small flat rhombohedra of a pale pink or rose color occur at DeTurk's, in Exeter township, four miles east of Reading. Dr Hoffman found the specific gravity=2.79, and the composition:

Carbonic acid	=	41.65
Manganous oxide	=	8.20
Ferrous	=	trace
Lime	=	44.36
Loss and water	=	5.79
		100.00

It is remarkable for the large quantity of carbonate of manganese which it contains=13.28 per cent.

Siderite. [158. *page* 159 *B.*]

A massive, slightly crystalline, gray siderite from Dallastown, York county, collected by Prof. Persifor Frazer, Jr., was at my request analyzed in the laboratory of the University of Pennsylvania, by Mr. Alfred Pearce, who found it to contain:

Carbonate of iron	=	77.99
Carbonate of manganese	=	0.45
Carbonate of magnesia	=	3.53
Carbonate of lime	=	1.43
Alumina	=	2.81
Quartz and silicic acid	=	11.56
Water, organic matter, &c	=	2.23
		100.00

Dr. Hoffman observed that at Maiden Creek and Tulpehocken, where the mine waters from the Schuylkill Coal Regions come in contact with lime, the river pebbles are thickly coated with a rusty crust of carbonate of iron and sulphate of lime.

Aragonite. [161. *page* 162 *B.*]

A white fibrous aragonite in seams and crystalline crusts in the waterlime has been observed by Mr. H. C. Lewis opposite Mt. Union, Mifflin Co., 2 miles E. of Matilda Furnace.

The analysis which I have made gave the following results:

Carbonate of lime	=	98.09
Carbonate of strontia	=	0.58
Silicic acid, ferric oxide, &c.	=	1.00
Water (by diff.)	=	0.33
		100.00

Strontianite.

Associated with the above aragonite Mr. Lewis found a mineral, heretofore not observed in Pennsylvania, which I proved by analysis to be strontianite. It occurs in minute crystals in acicular and divergent groups of a white color.

The analysis gave:

Carbonate of lime	=	15.36
Carbonate of strontia	=	83.15
Magnesia	=	0.38
Silicic acid, ferric oxide, &c.	=	0.50
Water (by diff.)	=	0.61
		100.00

Azurite. [170. *page* 167 *B.*]

Small seams of crystalline azurite of a deep blue color have been found near Reinhold's Station, in Lancaster county.

It has not yet been ascertained whether it occurs in sufficient quantities to be of value as a copper ore.

CHAPTER VI.

HYDRO-CARBON COMPOUNDS. [*pages* 196 *ff.*]

The importance of several of the natural gases for the manufacture of iron, &c., demanded the analyses of a few of these—which have been made by Prof. Sam'l P. Sadtler, of the University of Pennsylvania.

The following were analyzed, viz: that from the Burns Well, in Butler Co.; the Harvey Well, in Butler Co.; the Leechburg Well, in Westmoreland Co., and the Cherry Tree Well, in Indiana Co.

1. *Gas from the Burns Well.*

This well is located on the Delphy farm, ¾ of a mile from St. Joe, in Donegal Township, Butler County.

A qualitative analysis showed the presence of marsh-gas (CH_4), ethyl-hydride (C_2H_6), and propyl-hydride (C_3H_8) in traces; also the absence of the illuminating hydro-carbons (C_nH_{2n}).

Two quantitative analyses gave the following results:

Carbonic acid	=	0.34	—	0.35
Carbonic oxide	=	trace	—	trace
Hydrogen	=	6.06	—	6.15
Marsh-gas	=	75.75	—	75.12
Ethyl-hydride	=	17.84	—	18.39
Propyl-hydride	=	trace	—	trace
Illuminating hydrocarbons	=	——	—	——
Oxygen	=	——	—	——
Nitrogen	=	——	—	——
		99.99		100.01

The specific gravity of the gas (*moist*) was found=0.697. The specific gravity as calculated from the above analysis gave for the *dry* gas 0.6148.

2. *Gas from the Harvey Well.*

This gas is now utilized at the iron works of Graff, Bennett & Co., at Sharpsburg, and Spang, Chalfant & Co., at Etna, Allegheny county. It was collected from the supply-tubes at the works of the latter firm. The quantitative analysis gave:

Carbonic acid	=	0.66
Carbonic oxide	=	trace
Hydrogen	=	13.50
Marsh-gas	=	80.11
Ethyl-hydride	=	5.72
Propyl-hydride	=	—
Illuminating hydrocarbons	=	—
Oxygen	=	—
Nitrogen	=	—
		99.99

3. *Gas from the Leechburg Well.*

This well is across the Kiskiminetas river from Leechburg, and therefore in Westmoreland county. Qualitative tests were made, which, as in the case of the Burns well, showed the presence of marsh-gas, ethyl-hydride and a trace of propyl-hydride; the tests with bromine indicated traces of illuminating hydrocarbons. The quantitative analysis gave:

Carbonic acid	=	0.35
Carbonic oxide	=	0.26
Hydrogen	=	4.79
Marsh-gas	=	89.65
Ethyl-hydride	=	4.39
Propyl-hydride	=	trace
Illuminating hydrocarbons	=	0.56
Oxygen	=	—
Nitrogen	=	—
		100.00

4. *Gas from the Cherry Tree Spring.*

This gas issues from a strong spring of water at Cherry Tree, Indiana county. The quantitative analysis gave:

Carbonic acid	=	2.28
Carbonic oxide	=	—
Hydrogen	=	22.50
Marsh-gas	=	60.27
Ethyl-hydride	=	6.80
Illuminating hydrocarbons	=	—
Oxygen	=	0.83
Nitrogen	=	7.32
		100.00

APPENDIX TO HYDROCARBONS.

Lignite, [e page 179.]

Another locality, in addition to the two mentioned on page 180 B, is recorded by Prof. F. Prime, Jr., in his Report of Progress for 1875. He describes lignite as having been found in the clays of the brown hematite iron ore mine, at Ironton, Lehigh county.

INDEX

To Dr. Genths' report of 1875.

www.ingramcontent.com/pod-product-compliance
Lightning Source LLC
LaVergne TN
LVHW010254110826
845151LV00004B/1472

* 9 7 8 1 4 2 5 5 2 2 6 9 8 *